THE MEN WHO BUILT RAILWAYS

THE MEN WHO BUILT RAILWAYS

A Reprint of F. R. Conder's

Personal Recollections of English Engineers

Edited by Jack Simmons

Thomas Telford Limited, London
1983

First published by Hodder and Stoughton in 1868 under the title *Personal Recollections of English Engineers*
This edition published by Thomas Telford Ltd, 26–34 Old Street, London EC1 in 1983

ISBN 0 7277 0183 5

Typeset by Herts Typesetting Services Limited, Hertford
Printed and bound in Great Britain by Redwood Burn Limited, Trowbridge, Wiltshire

Editor's introduction

In spite of the efforts of biographers, we know little that is truly personal about the engineers who built and equipped the British railways. Most of them had a hard life—above all those who were at work during the principal age of construction, in the 1830s and 1840s. Not many lived to be old. They were men of action, and when action was past their work was over. They did not feel the need to justify themselves by setting down their recollections. George Stephenson, is is true, devoted much time in the closing years of his life to relating his own earlier history, at public celebrations and whenever he saw a chance to proclaim the moral lessons he drew from it. He also talked about the past, to Thomas Summerside and others. His son Robert conversed with Samuel Smiles and drew up his own account of the design of the Conway and Britannia Bridges.[1] But neither of them wrote reminiscences. Only one of the leading men of the early days did that: Daniel Gooch, who recorded the story of his own life as far as 1867. He also kept a diary, which he continued to 1885.[2] So did one other well-known railway engineer —Vignoles;[3] but not, as far as I know, any more.[4]

The *Personal recollections* reprinted here are of quite a different kind. They are the work of a man who left no mark on railway engineering but went after a time into the business of contracting. When he returned to engineering in later life he concerned himself with other things than railways. He was Francis Roubiliac Conder. Although he published the book anonymously in 1868 as the work of 'A Civil Engineer', he acknowledged it as his own in 1874.[5]

Conder was the son and grandson of London booksellers. His father Josiah had just moved into St Paul's Churchyard (then the headquarters of the trade) when he was born, on 26th November 1815. His mother was a granddaughter of the great sculptor Roubiliac: hence his second Christian name. He was educated at Mill Hill School. In 1819 Josiah Conder sold his bookselling business and moved out of London to devote himself to literature as a profession. From 1824 to 1839 he lived at Watford,[6] and this book opens with an amusing

[1]Printed in J. C. Jeaffreson, *Life of Robert Stephenson,* 1864, ii, 285-304.

[2]Sir D. Gooch, *Memoirs and diary,* ed. R. B. Wilson, 1972.

[3]The original is in the British Library: Add. MSS. 35071, 34528-36. The diary is a work of extraordinary detail, to have been written up daily by a civil engineer in extensive practice. It has been used in the good recent biography of him by K. H. Vignoles, 1982.

[4]There are diaries of some lesser railway engineers, such as that of R. B. Dockray, surviving in fragments between 1850 and 1860 (ed. M. Robbins, *Journal of transport history,* 1965-66, vii), and the young Henry Swinburne's of 1842-43 (Northumberland Record Office, ZSW 539/2, 3). *John Brunton's Book* comprises an interesting set of reminiscences, written down for its author's grandchildren and published 40 years after his death, in 1939.

[5]On the title page of his *Child's history of Jerusalem.*

[6]E. R. Conder, *Josiah Conder: a memoir,* 1857, 112, 118, 236.

i

sketch of that town as it then was, before the railway brought great changes to it.

At the age of eighteen the boy was taken on as an articled pupil of Charles Fox (less than six years older than he was himself), to work on the London and Birmingham Railway under Robert Stephenson.[7] We can piece together a rough outline of his successive railway employments from the narrative here. From the London and Birmingham he moved to the Eastern Counties Railway in 1835, and then he went to the Birmingham and Gloucester, about which he has a good deal to say that is interesting in chapters X-XIII. Another young assistant who served there with him was Herbert Spencer, who wrote his own account of the undertaking many years later, with which Conder's may be compared.[8] After that railway was opened in 1840 he worked for Charles Fox, in partnership with John Henderson as a contractor at Smethwick, outside Birmingham. His duties with the firm included the supervision of one of the most curious pieces of construction in these years—the running of a railway through the long tunnel of the Thames and Medway Canal in 1844-45. In 1848-55 he was engaged intermittently as a contractor for works on the South Wales Railway, under Brunel, in Pembrokeshire. He also went over to Ireland, to represent Fox Henderson on the construction of the Cork and Bandon Railway, which afforded wryly entertaining material for chapters XXII and XXIII of this book. He set up his own firm, Conder Goode & Co., and went to look for railway contracts abroad, securing one in south-western France and negotiating for another in Portugal that he did not get. It would have been well for him if he had not succeeded in a third bid he made—for the construction of part of the long line between Naples and Brindisi. That took him out to Italy in 1856, and there he was engaged for almost nine years[9] in an increasingly hopeless struggle, which cost him all the capital he had accumulated from his work at home and brought him to brain fever, nearly to death.

The risk in this enterprise must have been obvious from the start. Its success depended entirely on the goodwill of the government of Naples, a rigid autocracy endangered by the hatred of a large number of its people and notoriously faithless in all its dealings. Although it promised Conder support and encouraged his survey in Apulia, in fact it obstructed him and presently seized the sum of £50,000 that he had been required to lodge with it as a deposit at the beginning. The government itself was overturned in 1860 by Garibaldi—Conder saw his entry into Naples; but in spite of diplomatic support from England he recovered from the new Kingdom of Italy nothing of what he had lost. Late in 1864 he returned home, ruined.

He had 30 years of engineering and contracting behind him, and

[7] The main source of information about Conder's life is his obituary in the *Proceedings of the Institution of Civil Engineers*, 1890, c, 379-383. It does not entirely agree with what he himself tells us in this book.

[8] *Autobiography*, 1904. He mentions Conder twice (i, 131, 138), recording a remark made to him by Charles Fox that "he [Conder] has not got his wits about him nearly as much as you have."

[9] The dates of his Italian employment are given in his book *The trinity of Italy*, xi-xii.

he may well have hoped to pick up the threads of it again quickly, for the third and last wave of feverish railway promotion was now rising in Britain. But he had been out of the way for a long time. Before he could get far in his quest for work the crash of May 1866 arrived, and all railway promotion was reduced or stopped. For the moment he turned his energies to writing, to produce in quick succession two substantial books: *The trinity of Italy* and this one.

Englishmen then took a great interest in Italian politics, and it was a good idea for Conder to turn his experiences there to account. *The trinity of Italy* is a vivacious piece of work, and was well received on its publication in 1867. By that time railways were much in the news, and Conder decided to record in the same way something of what he had seen in his earlier life. He wrote quickly, and these *Personal recollections* came out in 1868.

The two books must have helped to make a place for him in high-class journalism, on which he came to live in his later years. Nearly all that he published here was anonymous. In 1872–88 he wrote much for the *Edinburgh Review,* as well as for *Fraser's* and the *Art Journal;* he commented on politics for the *St James's Gazette* and for some time had a weekly article in the *Builder.* The books he produced—slighter than the two of 1867–68—reflected an old interest in philosophy[10] and a new one in the history and archaeology of Palestine, stimulated by the work of his son Claude Reignier Conder, who was at the beginning of a distinguished career in that field.[11]

In the last decade of his life he spoke out, almost for the first time openly in his own name, on the place that he thought canals could, and should, take in the commercial life of the country. He supported with zest the case for the Manchester Ship Canal with two publications in 1882: a *Report on the comparative cost of transport by railway and canal* and *The actual and the possible cost of conveyance between Manchester and Liverpool,* in which he argued that that cost was then higher than it had been in 1829, before the opening of the Liverpool and Manchester Railway.[12] In the following year he gave evidence to a Commons Committee on Canals, designed to show what might be done with them in England from Continental examples;[13] we can detect in his questioners a certain impatience with this elderly witness, who was obliged to admit that he had never had anything to do with the making or management of canals in his life. But his work was not confined to writing and talking. He had by now found an active and successful role as a sanitary engineer, making himself a recognised authority on the deodorisation of sewage.[14] He died at Guildford from angina, quietly reading in his chair, on 18th December 1889.[15] He never recovered the fortune he lost in Italy. His will was proved at £534.

[10]In 1850 he had produced *Elements of Catholic philosophy, or Theory of the natural history of the human mind.* He followed this up with a short work in German, published at Leipzig in 1879: *Drei Ideale menschlicher Vollkommenheit.*

[11]The Conders were an interesting family, with a diversity of interests. One of Francis's nephews was Charles Conder, the painter (1868-1909).

[12]Copies of these pamphlets are to be found in the library of the Institution of Civil Engineers: Tracts, vol. 350.

[13]*Parliamentary papers,* 1883, xii, 127-144.

[14]A system devised by him was installed at Windsor Castle.

[15]Very brief obituary in *The Times,* 19th December 1889.

Conder's life seems strangely fragmented. His interests were wide, diverse and perhaps scarcely consistent with one another. The religious strain in him was strong and stoutly Protestant. Southey had written of his father that he held 'most of the opinions which were in fashion under Cromwell—a thorough Independent'.[16] Francis Conder was vehemently opposed to the Roman Catholic church, to ultramontanism and to the claims of the Popes to universal dominion. He looked back with relish over 'the twenty-nine schisms which have rent their communion'.[17] His mistrust of Catholics appears in this book, and of the Tractarians who seemed to him Catholics in disguise (pp 111). That dislike was doubtless sharpened by his experiences in Italy, in the realm of the most bigoted Catholic monarch in Europe, Ferdinand II of Naples. And yet he gives a not wholly unfavourable account of that ruler, rather different from the portrait of the cruel and sinister King Bomba that was accepted by conventional British Protestants and Liberals.

It is quite plain, both from this book and from all the other evidence, that Conder did not succeed in railway work. With his varied experience and obviously quick and intelligent mind he would have been a well-qualified candidate for at least a senior assistant's post, as Deputy or Resident Engineer, in the 1840s when such men were most in demand, during the great boom in projection. But he got nothing of the kind, and turned to contracting instead. Compare him with William Baker, his junior by a year or two, who joined the railway service with him (p. 11). In 1852 Baker became engineer of the Southern Division of the largest railway in the country, the London and North Western. Seven years later he succeeded Robert Stephenson as engineer of that company's whole system. By the side of this former colleague, and of other contemporaries, Conder must have seen himself as a failure. When he returned to England at the end of 1864, poor and in need of work, it does not appear that even his old chief Sir Charles Fox found anything for him to do. He published these recollections not only in the hope of making money by them but also in some degree—it is evident—in order to justify himself, to explain failure away.

And yet they are not embittered. He too, like his father, was a thorough independent—a man who delighted in forming opinions for himself, shaded exactly according to his own principles. That is one of the things that stands out here. Scarcely anybody who is described in these reminiscences, whether by name or indirectly, is accorded Conder's unqualified admiration. But equally, his most unfavourable accounts of the men he disapproved of or disliked nearly all include some recognition of what he thought their good qualities.

iv   Conder kept his own name most strictly out of the book. His rule,

[16] Selections from the Letters of Robert Southey, ed. J. W. Warter, 1856, iii, 275.

[17] The trinity of Italy, xviii.

as he explains at the beginning, was to put names to the men he praised but to withhold them from those he attacked. He names the most distinguished: Robert Stephenson and Brunel, the Irish contractor William Dargan, the Government Inspector Sir Charles Pasley. Nearly all the rest of the figures in the book are anonymous, and—it must be said straight away—many of them elude identification now. This practice is not uncommon in memoirs, but it has seldom been pursued so thoroughly by anyone as by Conder. He enjoyed mystery for itself—the mystery of his own authorship first of all. That does not destroy the value of his book. In one sense the method he pursued may even increase it: for it allowed him to record exactly what he wished to say, however unfavourable, without the need to pay much attention to the law of libel. By the time he wrote, most of the leading men he spoke of were dead, and so beyond invoking that law;[18] but it would have been impossible for him to trace the smaller fry, and he covered them all with a blanket of anonymity. These men survive in his pages as types, sometimes resembling others we meet elsewhere. The types are not fictional, however, for Conder persuades us to accept them as portraits sketched from the life.

[18]The chief exceptions, I think, are John Braithwaite and Sir Charles Fox, who did not die until 1870 and 1874 respectively.

At the head of those he attacks stands W. S. Moorsom, the engineer of the Birmingham and Gloucester Railway: 'a man too full of contradictions to be dismissed in a line', he remarks; 'perhaps difficult to be judged with fairness' (p. 78).[19] He allows him some personal virtues but he sees those as quite cancelled out by three defects. In the first place, Moorsom came to the full responsibilities of a chief engineer without any adequate experience. There was truth in that. He had been in the army, mainly in Nova Scotia, until 1832 and had been brought in to work on the London and Birmingham Railway at the instance of his elder brother, who was a director of the company. There he attracted Robert Stephenson's notice by a meritorious piece of surveying, and he was set on his new path. He sensibly tried to repair his deficiencies by an extended tour of the most important English railways in 1835–36, but when the Birmingham and Gloucester company appointed him he knew very little of the duties the post required. Although that was a serious deficiency, as Conder says it could have been remedied if he had chosen for his assistants men who had the knowledge and experience that he himself lacked and had been willing to trust them. The second charge made here is that he did nothing of the kind: 'his professional knowledge was limited, to use the most favourable term; but his chief hostility was evinced towards those who, in perfect good faith, would have enabled him to supply the defect' (p. 78). Here were two grave faults. In Conder's eyes Moorsom had a third, much graver still. He was

[19]It is worth noting that in Moorsom's case Conder comes near to departing from his rule of anonymity, calling him "Captain Transom" (p. 85).

v

devious, shifty, and unreliable. 'He managed to teach all his staff that they would meet fairer dealing from anyone than from himself . . . He was more likely to say nothing *to* the defaulter, but much *of* the default where it might most turn to his disadvantage' (pp 78–79).

How far are we to believe these charges? Conder was opinionated. Did he quarrel with Moorsom, and were these statements made in revenge? We have no evidence directly on that point, but we can test Conder's assertions on another, linked with it. He observes that Moorsom had been 'borne well forward on the flow of the first railway mania' (i.e. in 1835, when he secured his appointment on the Birmingham and Gloucester Railway) and was therefore 'on the very crest of the wave' in the second, ten years later. 'He took into Parliament more railway Bills than almost any other engineer. He lost them all but one, and that one was secured by the exertions of the experienced contractor who intended to make the line' (p. 87). Those statements are almost exactly true.[20] The one English railway that he carried through successfully was the Southampton and Dorchester, and the contractor there was Peto.[21] To that must be added the Waterford and Kilkenny Railway in Ireland. However, the survey there, on which the authorising Act was based, seems to have been deeply discreditable to him, and the work of construction was quickly taken out of his hands.[22] He then almost disappeared from the roll of practising railway engineers.[23] He never constructed any other line except the Ringwood Christchurch and Bournemouth, eight miles long, in 1860–62. That was at the end of his life; he died in 1863.

Reviewing Moorsom's career as a whole, it must certainly be judged a failure. He was not incompetent. The erection of the Birmingham and Gloucester's cast-iron bridge over the Avon at Defford won him a Telford medal from the Institution of Civil Engineers. In 1850 he gained first prize, against 60 competitors, for his plan for a road and railway bridge over the Rhine at Cologne. But his work was seldom satisfactory. The Lickey incline (at 1 in 38), which the Birmingham and Gloucester Railway adopted on his advice in preference to more circuitous routes recommended by George Stephenson and Brunel,[24] was a serious mistake. It is to be noted that both the Cornwall and the West Cornwall companies' lines were taken away from him and given to Brunel to revise and execute.[25] Conder's judgement on Moorsom's public performance as an engineer may be taken as well-founded and reliable. It is harder to assess the truth of what he says about Moorsom's character. But the extremely discreet Herbert Spencer goes some way towards confirming the unfavourable opinion of him that is expressed here.[26]

Briefer glimpses of other well-known engineers are also given us in this book. Braithwaite, Vignoles, and Rastrick are set down, all

[20]Reading between the lines, we can see that they are borne out even in the very circumspect obituary of Moorsom in the *Proceedings of the Institution of Civil Engineers,* 1863-64, xxiii, 498-504 (especially 501).

[21]See J. C. Cox, *Castleman's Corkscrew* (1975).

[22]The Government Inspector's report on the first section of it to be finished (10th September 1847: *Parliamentary Papers* 1847-48, xxvi, 438) says: 'the Parliamentary section . . . according to the statement of the present engineer [Charles Tarrant] . . . is completely wrong, the surface of the ground as it exists being entirely of a different form from that delineated on the Parliamentary section'. Compare that with the sentence in which Conder describes the care taken by Brunel to secure a correct and economical survey of the South Wales line (p. 85).

[23]He was engineer to the Cromford and High Peak Railway for a time and began its modernisation; but the company found itself too poor to retain his services, and he was dismissed in 1856: J. D. Marshall, *The Cromford and High Peak Railway,* 1982, 12-13.

[24]*Proceedings of the Institution of Civil Engineers,* 1864, xxiii, 500.

[25]The cracks were papered over. On the Cornwall Railway the change was ascribed to Moorsom's excessively large commitments to other railways. On the West Cornwall he was personally acceptable; he served as chairman of the company in 1847-50.

[26]*Autobiography,* i, 183-185.

alike, as 'hard swearers' (pp 36, 154). Towards Rastrick Conder is unjust in saying that he 'came to the front of his profession rather in virtue of Parliamentary tactics than by any other form of introduction' (p. 154). No doubt this remark applied to his skill as a witness before Parliamentary committees, but it ignores his work as engineer of the main system of the London Brighton and South Coast Railway[27] as well as his important part in the development of the steam locomotive.[28] The Gravesend and Rochester line, discussed in this book, was one of his last undertakings, carried through when he was 64. He may well have grown testy by that time. (Indeed if we were to judge solely by Conder's stories, it would seem surprising that any railway engineer managed to avoid an early death from apoplexy.) But the account of this railway given here finishes with a smiling portrait of Rastrick—named at last—with 'his face and eyes glowing under his snowy hair', pouring out reminiscences at a fish dinner to celebrate the successful inspection of the line. The Inspector was Sir Charles Pasley, and Conder's accounts of him are vivid and memorable (pp 142, 154). His physical courage and his sense of humour stand out in them, as well as (from the point of view of a railway engineer) the limitations of his knowledge, and his bureaucrat's timidity.[29]

The account of William Dargan, the Irish railway contractor, in this book (p. 165) is entirely laudatory. He appears as a just man, both in his business dealings and as an employer of labour: decently modest, and a notable benefactor of his country—a verdict that, broadly speaking, stands. The one thing noted as regrettable is that Dargan shortened his life with 'Irish sherry'.

Conder is at his best on Robert Stephenson and Brunel, expressing his admiration for them and recording quite fairly what he sees as their defects. He brings out Stephenson's energy, his endurance, his extraordinary strength, his honourable dealing, together with the autocratic tendency of his temper and his fieriness when crossed. His modern biographers have recognised the merit of Conder's account of him. These are 'the words of an admirer but not a worshipper', says Michael Robbins; 'and they ring true'.[30]

The treatment of Brunel here is more complex and quite as interesting. It may even be better because it is based on more mature and, as he says himself, more intimate observation. The first thing that struck him was Brunel's insistence on keeping control himself of the whole of the enterprises that were going on under him, compared with Stephenson's devolution of authority to his subordinates;[31] the second, Brunel's 'extreme and unprecedented insistence on excellence of work' (p. 119). He is surely right in saying that that insistence added substantially to the expense of building Brunel's railways, not only because the engineer required a high and therefore costly

[27]He executed the main Brighton line to Sir John Rennie's plan. He was sole engineer for the lines from Brighton to Chichester and Hastings.

[28]His notebooks in the Goldsmiths' Library of the University of London show what a serious student he had been of both the civil and the mechanical engineering of railways.

[29]Conder's account can be checked by Pasley's own. His reports on three inspections of this line and its tunnel in 1844-45 (they mention Conder) are in *Parliamentary Papers*, 1846, xxxix, 478-480, 502-504.

[30]*George and Robert Stephenson*, 1981 edn, 44.

[31]This is borne out in ways of which Conder himself cannot have been aware. Brunel even conducted some of the negotiations for the purchase of land for his railway companies himself. See, for example, his dealings with Lady Georgiana Fane's solicitor concerning Brympton d'Evercy (Somerset Record Office, DD/BT 14/3/1) and with Lord Vivian's concerning Glynn (Cornwall Record Office, AD 267/90).

standard of materials and workmanship but also because contractors came to tender for his undertakings only at top prices (pp 122–123).[32]

[32]Brunel did not disregard the need for economy, however. The cost of the Royal Albert Bridge at Saltash for example (£225,000) was extraordinarily low: *The Works of Isambard Kingdom Brunel,* ed. Sir A. Pugsley, 1980, 178.

Brunel was a perfectionist, for whom the best might be the enemy of the good. Many people have said that, in admiration or criticism, from his day to ours. But Conder does not leave it there. His professional experience enabled him to take the measure of Brunel, not of his art and intellect alone but also of the science and observation on which they were founded. He spells it out plainly for the lay reader. A difficulty arises, for example, from a slip in the wall of the harbour he is building for the railway at Neyland in Pembrokeshire. Conder describes for us in words, with almost the accuracy of the best television, the process of observation and thought by which Brunel solves his problem (pp 129–130). In the end, with no disloyalty to Robert Stephenson, Brunel stands out for him as 'our greatest engineer' (p. 187).

Conder's book bore exactly the right title. His recollections are *personal:* personal to the men he talks about, personal to himself. They are therefore individual, sometimes prejudiced. By the time he recorded them he was out of the railway world and could say what he chose with the fullest freedom because he wrote anonymously. They were *recollections,* of things he had observed from 12 to 35 years before he published them. Their interest for us is not confined to what he has to say about railway work. His eighth chapter is one of the best accounts we have of the pleasures of travelling by coach, free of commonplace romanticism and carrying with it entire conviction. The weaknesses of the book are obvious: the temptations to malice, to a little paying off of old scores (from which the author was by no means free), the insidious tricks of memory. But there is strength too. Conder was able to write with detachment. He could look back coolly on his earlier experiences, reflect on them and place them in perspective. At the same time these recollections were not, like so many others, set down in old age. The writer was in his early fifties when he recorded them, with a fully alert mind ready to seize new opportunities as they came his way. The succeeding 20 years of his life show that.

For us now, more than a hundred years later, the book represents his greatest stroke of luck: the chance to set down a first-hand account of railway construction in Britain at the time of its fullest intensity. Our knowledge of that story is already considerable, from official records of many kinds; and what they have to tell us is far from exhausted yet. Those records are dry and formal. If some of the biographies of the leading men involved, written shortly after they died, are good, they were inhibited from entirely plain speaking. Conder was not himself a leading man, but he served those who were, and he watched them closely. Although his book is full of anecdotes, they are not—like so many others, handed down from one writer to another—stale: some of them are misleading, some difficult to interpret. Conder made no claim to be a judicious historian. He was writing reminiscences. They come out fresh and alive here, recorded with spirit by a man who knew very well what he was writing about, and who also knew how to write. They have been almost entirely forgotten. They deserve to be rediscovered now.

This reprint follows the edition of 1868 exactly, with the correction of a few printing mistakes. There were no illustrations. Those that have been inserted here are from contemporary engravings.

Illustrations

Contents

PRELIMINARY CHAPTER

THE AUTHOR TO THE READER

The main part of the following pages had been written before the author had the opportunity of perusing those very interesting "Lives of Engineers," which were published during a long absence on his part from this country.[1] Before placing the manuscript in the hands of the publishers, however, it became necessary to read the volumes of Mr. Smiles, in order to see whether there had been any unintentional interference with the manor of a gentleman who had so ably vindicated his right to be considered an historian of Engineering.

In addition to the pleasure and advantage of the perusal of the works in question, the author of the following sketches arrived at the conviction, that no possible conflict could occur between publications of such a different character. The volumes of Mr. Smiles are literary and historic. Their author has had recourse to the elements of narration that lay ready to the diligent hand, but he has not written either as an eye-witness, or as an Engineer.

The following recollections, on the contrary, are almost exclusively original and personal. They come now, for the first time, before the public. No incident, so far as the writer is aware, has been before published, with the exception of the newspaper reports of some of the Irish trials. It is certain that the greater part, if not the whole, of the matter must be new. And it is hoped that it will not be found altogether devoid of interest.

It has been the wish of the writer to avoid giving pain to anyone. The greater number of the men of whom mention is made in these pages, are now beyond the reach of praise or of attack. In those instances in which a personal anecdote tends to raise the estimate already formed of an eminent man, names have been frankly given. In those cases in which a regard for truth has compelled the author to draw the unfavourable features of a portrait, not only has the name been withheld, but it is the hope, as it has been the effort, of the writer, that such reticence has been used, that the feelings of no private friend or relative can be hurt, and that those only shall recognise the truth of the narrative, who, having themselves

[1] This book by Samuel Smiles appeared in 1861-62. His *Life of George Stephenson* had been published in 1857.

1

had professional acquaintance with the subjects, will fully coincide with what has been said. Even of those old brother subalterns, alas! the number now left is very small.

Death has reaped a rich harvest among the men, who came to the fore when we entered on the construction of our railway system. Robert Stephenson and his contemporaries died, full of honour and not unrewarded by wealth, but barren as to length of days. The iron frame and noble intelligence of Isambard Kingdom Brunel succumbed to his contempt for the laws of health. He crowded the activity of a long life into a few short years, and his unresting devotion to his profession, made him a victim, in middle manhood, to the infirmity of age. Very few men are now among us who entered the profession before the period when the writer was an articled pupil on the London and Birmingham Railway.

At that time a new feature was becoming apparent in English literature. In the columns of the *Morning Chronicle* appeared, from time to time, articles with the signature of ''Boz,''[2] which gave descriptions of scenes and grades of English society which were new, except for incidental reference, to the writer of fiction. The taste of the public welcomed the new essay, and a light periodical literature has been the result, which has accustomed the majority of readers to the admiration of a sort of Dutch painting in words, and to the preference of descriptions of what is really actual and lifelike, however humble, to that which was formerly thought the nobler province of fiction.

From this newly originated form of chronicle, the chosen ground of the novelists of the second portion of the nineteenth century, the scenes witnessed by those who introduced our existing railways into the districts once traversed by well-appointed coaches, have hitherto escaped. Engineers and contractors have made their appearance in the pages of the novelist,[3] but they have been, for the most part, the offspring of imagination, and not the transcript of actual life. And as all fiction grasps the mind only in proportion to its resemblance to fact, it may chance that, to many readers, incidents of actual occurrence at a period of social transition, and records of a state of society strange to those who have grown up since England was bound together by the iron net of railways, will not be without a certain charm. Nor is the object of the author exclusively to amuse. It is his hope that some information, not altogether worthless, and some results of reflection, not altogether without value, may be found in the following pages.

The very flattering reception which the reviewers, almost without a single exception, have given to the author's *Italian Reminiscences*, has emboldened him now to offer to the public his recollections of

English Engineers. It is a grateful task to refer to a first reception,

[2]The first of Dickens's articles was published in 1833; *Sketches by Boz* appeared in book form in 1836-37.

[3]Daniel Doyce, for example, engineer and inventor, in Dickens's *Little Dorrit,* 1857; Sir Roger Scatcherd, contractor, in Trollope's *Doctor Thorne,* 1858.

which has encouraged a second appearance; and it is the hope of the writer, that anecdotes of Stephenson and of Brunel, of engineering, and of railway warfare and advance, and of the revolution effected by the introduction of railways into England, Wales, and Ireland, will not be less attractive than tales of Ferdinand and of Francis of Naples, of Garibaldi and the English legion, and of the struggle between civilization and the Papacy.

It is proper to state that the Railway Regulation Act of 1868, which comes into full operation on 1st April, 1869, empowers the Board of Trade to authorize the Construction of Light Railways, on the two sole conditions, that the maximum weight to be borne on a pair of wheels shall be eight tons, and the maximum speed, twenty-five miles per hour. But this provision, unless attended by a change in standing orders, does not authorize the application to Parliament for power to construct a railway, originally intended to be "Light." It is a boon to existing Companies, but it is only an admission of the fact that such a boon ought to be extended to the public.[4]

[4]See note 111.

November 16, 1868.

CHAPTER I

A country town before the railway era

The nineteenth century had just run one-third of its course. The reigning county belle was beginning to think it inconvenient that she had been born with the century, so that the date *anno domini* recalled to all who were in the secret her exact age; not that she had any good reason for displeasure, for she had already made two ventures in the lottery of matrimony, and on the second occasion had secured a high prize,— as far at all events as fortune was concerned,— for incomes of two thousand a year were less common at that time, and in that part of the world, than may now be the case. The excitement caused by the debating and by the passing of the Reform Bill, and by the county elections connected with the contest had subsided, after having culminated in an unprecedented exhibition of rustic sports, including jumping in sacks, climbing a soapy pole, and catching a pig by the well-soaped tail, and in a display of fireworks suited to the gravity of the occasion. The little country town in question had therefore subsided into its usual state of dull and uneventful repose. It was not a spot possessing in itself any peculiar charms, although it lay in the neighbourhood of three of the finest parks that adorn the island.[5] The route of the Birmingham mail, crossing the river and mill course by two bridges, ascended a long straight street, the chief peculiarity of which was, that about every third house seemed to bear the sign of a tavern, hotel, or ale house. The church boasted a large shapeless tower, disfigured by roughcast, from the centre of which arose a ridiculously small and disproportionate spire. The Baptist meeting-house, after the modest fashion of former days, was hidden amid a group of poorer houses, at some distance from what was dignified by the name of the ''High Street,'' and was only accessible by three lanes or alleys, which, in wet and muddy weather, were so emulous of discomfort as to be about equally impassable. But the freedom of private judgment was evinced by a third edifice, about the size of a good-sized room, but built in the form of the Noah's ark of the toy shops, which bore on its diminutive pediment a designation, implying that its fre-

[5]The town is Watford. The 'three parks' are The Grove, Cashiobury, and either The Lodge or (more distant) Moor Park.

5

quenters were the sole exceptions to the general immorality and blindness of their neighbours. The high priest of this self-contented company, at the time to which we refer, exercised, six days in the week, the art of a shoemaker, and announced himself on his signboard as "Fitall, from London." I hope his sermons were better than his boots, for I have yet a painful remembrance of being persuaded by his eloquence into being measured for a first pair of Wellingtons, and being further persuaded, when they turned out to be at the same time much too large and much too small, to persevere in wearing them, the result of which was an injury to the instep, that interfered with comfort for a considerable time. Passing the Antinomian ark, passing the large corn mill recently rebuilt, (out of which two dusty miller's men were constantly peering or passing, and around which gambolled a large flock of tumbler pigeons), running up by Wheatsheaf, Compasses, Odd Fellows' Arms, Leathersellers' Arms, Rose and Crown, George, and innumerable other signs, passing the long, low market-house, room above, and open pillared space below, leaving to the left the shady gardens of the vicarage, and the adjoining churchyard elms, you came to a few comfortable houses of respectable proportions, standing in their own gardens and grounds, all adorned by fine timber, and then at a cross road leading from the abbey and town of the first English martyr, to a noble park where once

⁶Cashiobury was a possession of St Albans Abbey.

stood a succursal of the monastery,[6] the town came to an end. The inhabitants of the spot troubled themselves but little with the proceedings of their neighbours, for then, and even to the present time, they kept themselves entirely clear of the perturbation caused by inviting lodgers from the metropolis. The hours of the single mail coach, that ran through the town, were so inconveniently early and late, that it rarely took a local passenger. A stage coach ran from a country town a few miles further, and this, with the horses and vehicles of the resident gentry and tradesfolk, fully supplied the requirements of the place. Two of the magnates of the place kept chariots, a gig or tilbury was the fashion of the generality of those who drove. The two or three religious communities—for there were rumours of the existence of a "Methodist interest," located in a yet more secluded spot than their Baptist rivals—lived in a peaceful state of non-intercourse with one another, the wrath of the honourable

⁷The Hon. W. R. Capell, Vicar of Watford from 1799 to 1855.

and reverand vicar,[7] a parson of the fast old country stamp, being reserved for those whom he designated as "devil-dodgers," that is, persons who went to church in the morning, and to chapel in the evening. Under this discountenance, devil-dodging was a very rare offence. The vicar was a figure not readily to be forgotten, not easily to be realized by the rising generation. A tall, hale, portly man, who being older than the century, had come into some trouble

6

through taking the advice of his hairdressers. Whether it was, as he said, that his hair was too red, or whether it was with a view to obliterate any of the foot-prints of time, his reverence had made experiments on his locks, which resulted in his appearance one Sunday in the reading desk, with hair of a brilliant green, and on a subsequent occasion, of a lovely purple. He rode very straight across country, but out of a conventional respect for the cloth, usually hunted in pepper-and-salt. In fact he discredited the tale,—that on one occasion a funeral had been for some time awaiting his ministrations, when he rode up to the churchyard with the dogs, put the surplice on over his scarlet coat, committed the brother to the dust, threw off the surplice, and harkaway after Reynard,—by the remark, that it *must* be a lie, as he always wore a mixture. To cards he was extremely devoted, although with only ordinary luck. In other respects he had been wild, but then, as he always observed, every apparent inconsistency between his habits and his profession was the fault of his father, who put him into the church, for which he had no inclination, merely to fill the family living; and what else could he do? So he made a point, during the London season, of returning from his noble sister-in-law's Saturday evening card table in good time for the morning service on the Sunday; and if the coach was late, or he failed to secure a promised lift, or in any way a little exceeded the limits of punctuality, the parish clerk, a congenial spirit, would aid him by putting back the church clock. The honourable vicar was generally accompanied by one or two very large dogs, one of which has been known to provoke irreverent remarks from the congregation by making his appearance in the pulpit, while his master was in the reading desk below, and placing his paws upon the cushion. He occupied much of the time spent on his cure, as he grew less fond of hunting, in attending to a fishery, by means of which he contrived to establish a feud with most of the mill-owners on the river; and where the curious might admire a minnow tank ingeniously constructed out of six tombstones, removed, surreptitiously it was thought, from the churchyard. For the rest, the Honourable and Reverend, as he was usually called, employed an agent to collect his tithes and Easter dues. He once threatened an importunate tradesman to get himself made a bankrupt *as a fishmonger.* His life was consistent to its close, and it was always said of him, that he was no one's enemy but his own.

There had been an exciting conflict between this honourable vicar and his diocesan. The bishop, rather a reforming man for those days, recommended him to keep a curate. The vicar really could not afford it. The bishop insisted, with no better result. The bishop nominated a curate and sent him down to the parish. The vicar stood siege,   7 stood in the breach on every occasion, and resolutely kept both bishop and curate at bay. On one occasion, having trusted rather too implicitly to the operations of his ally the clerk, on the face of the church clock, the vicar was so far behind time, on a Sunday morning, that the intrusive curate entered the vestry, indued the surplice, and was actually walking to the reading desk, when the former arriving in the very nick of time, popped into the desk unrobed, began the service at once, and sending an attendant for the surplice, put it on then and there as he continued his ministration.

With all that was wild and quaint about the character of this reverend gentleman, he knew how to assert his own position, on occasions, with adroitness, as well as with dignity. Unlike as he may be to the clergy of a later generation, he was but a *remanet* of those for which the county had formerly been notorious. The Evangelical clergyman who was appointed about this time to an adjoining rectory, in the gift of one of the Oxford colleges, was said to have been the first sincere and earnest preacher beneath its roof since the Reformation. So that between the example of the vicar, and the thick growth of houses of refreshment, it was not matter of surprise that the Sunday was rather honoured in the breach than in the observance in the town. Some of the graver inhabitants saw room for much improvement. A well-known literary character who resided in the neighbourhood, a man loved by all who knew him, not more for his facile pen and his ready rhetoric as a public speaker, than for the blameless purity of his life,[8] took part in the little reformatory movement; and a public meeting was convened, at which a neighbouring nobleman, who was also a Count of the Kingdom of Prussia, and held the ancient though obsolete dignity of Chief-Justice in Eyre,[9] consented to preside. The room was filled, and proceedings were about to commence, when general astonishment was caused by the unexpected appearance of the vicar. The conveners of the meeting looked at one another, expecting some grotesque scene, or some serious opposition. The audience looked like a flock of sheep startled by the sudden presence of the wolf. The Chief-Justice himself looked posed. All were astray but the vicar, who coolly advanced to the head of the table, and commenced an address to his lordship and his parishioners. ''On such an occasion as the present,'' said he, ''in which I am naturally so deeply interested, I might have ground to quarrel with my good friends who have brought you together, for supposing that I would allow anyone but myself to take the chair, and to give all the support of my position to your very laudable object. The only person to whom I could cede my right, has, however, most kindly condescended to come among us; and I have very much pleasure in moving that the Right Honorable the Earl takes the chair.''

[8]Probably Dr Thomas Monro (1759-1833), the patron of Turner and Girtin. He lived at Bushey.

[9]The third Earl of Clarendon.

The motion was carried by acclamation, and the vicar at once held his own, and put everyone in good humour. It had so chanced that it had never occurred to the most sanguine of the reformers to expect anything but opposition from the Honourable and Reverend, who thus ably gave a lesson in etiquette to the excellent nobleman, whom he privately and irreverently denominated ''Old Pickaxe,'' adding that he would have been better employed praying among his cucumber frames with his own gardeners, than interfering in the parish. So the conveners of the meeting were not inexcusable in their omission.

Through this tranquil and old-fashioned spot, at the date above indicated, rumour spread the intelligence that a railway was about to be constructed from London to Birmingham. What a railway was, and by what new and unheard-of evils it would destroy the repose of the country, no resident had any distinct idea. Some of the readers of the two or three copies of the *Times* that passed through the post-office were aware, rather as a matter of abstract historic announcement than of local interest, that in the preceding year a bill for this purpose had been thrown out by the House of Lords, in consequence of the opposition of a noble Earl, half-brother to the vicar,[10] through whose park Robert Stephenson, following the line selected by Telford for the Grand Junction Canal, had proposed to lead the railway. A second bill had been brought forward by which, at the expense of heavy works, the park and ornamental grounds of the Earl had been avoided, the line crossing the valley of the Colne by an embankment of what was, in those days, unprecedented magnitude, and thus boring beneath the woods, at a distance from the mansion, by an equally unprecedented tunnel of nearly a mile in length. These matters, however, only concerned the professional advisers of his lordship, and the announcement of the speedy commencement of the railway in the neighbourhood, was received by most persons merely with that sort of uninterested curiosity with which we now regard some unusually enigmatic placard, by which the art or science of the bill-sticker endeavours to call the jaded attention to the announcement that is *about* to be made.

At last one or two strange faces appeared in the town, and men in leathern leggings, dragging a long chain, and attended by one or two country labourers armed with bill-hooks, were remarked as trespassing in the most unwarrantable manner over pasture land, standing crops, copse and cover; actually cutting gaps in the hedges, through which they climbed and dragged the land-chain. Then would follow another intruder, bearing a telescope set on three legs, which he erected with the most perfect coolness wherever he thought fit, peering through it at a long white staff, marked with unintelligible hieroglyphics, which was borne by another labourer, and moved

[10]The fifth Earl of Essex.

9

or held stationary in accordance with a mysterious code of telegraphic signals made by the hand.

The farmers, naturally indignant, ordered these intruders from their fields. The engineers, for such they were, took but little notice. The farmers proceeded to threats. The ringleader of the invaders produced a red book, folded in an oblong form, from the voluminous pocket of his velveteen jacket, and offered it to the irate farmer as a sedative, informing him that it was the Act of Parliament by authority of which he was acting, and further specifying that it was a King's printer's copy of the Act, and therefore was legal evidence. The farmers, at their wits' ends, did all in their power to rout and baffle the intruders, but in vain. One thing alone remained for them to do,—and often they positively swore that they would have recourse to that extreme step—they would shoot the intruders. But the latter calmly replied that that was no business of theirs, and the farmers did not draw trigger. The blank look of petrified astonishment which crept over the furious face of an irate farmer, when his threat of instant death, enforced by the production of a weapon, was met by the calm rejoinder that the result was his look-out, and not that of his tormentor, would have been a fine study for a great comedian. It was so ridiculous as to be almost sublime.

The next step in the invasion proved a yet further aggravation to the farmers, although it was one which, for the first time in the course of the contest, afforded them the pleasure of retaliation. Loads of oak pegs, accurately squared, planed, and pointed, were driven to the fields, and the course of the intended railway was marked out by driving two of these pegs, one left standing about four inches above the surface to indicate position, and a smaller one driven lower to the ground a few inches off on which to take the level, at every interval of twenty-two yards. It is obvious that the operations of farming afforded many an opportunity for an unfriendly blow at these pegs. Ploughs and harrows had a remarkable tendency to become entangled in them; cart wheels ran foul of them; sometimes they disappeared altogether. A mute and irregular warfare on the subject of the pegs was generally protracted until the last outrage was perpetrated by the agents of the company; the land was purchased for the railway.

Flying parties of surveyors were now succeeded by the regular staff. A tall, very tall young man, upwards of six feet high, though losing somewhat from a slight stoop and a low-crowned hat,[11] was found to have actually rented one of the few available houses in the town. He was a man whom no one set eyes on without wishing to see more of him. A grave face, with a sweet and yet dignified expression, very dark eyes, lineaments such as those to be found

[11](Sir) Charles Fox (1810-74).

in the drawings of Westall, a forehead not high, but broader than any often met with in portraiture, in sculpture, or in life; the dress of a decent mechanic, the air of an educated and well-bred man, and no gloves: these were some of the outward marks of a man who has since made his mark in the country. He was one of a family, in a northern English county, distinguished for the talent of its members. He was educated as a surgeon, but on coming of age declined to follow his paternal profession, and, after an engagement under Ericsson, the inventor of the Monitors, during which he had a share in conducting those experiments on the Liverpool and Manchester Railway in which the speed attained by the locomotives so much exceeded the expectations of their constructors, became one of the earliest subalterns of Robert Stephenson. The colleague and senior officer of this engineer was a man considerably older, and rather of the stamp of the old country surveyor, or the engineers of the school of Telford, than of a mechanical turn.[12] His shrewd grey eye, half inquisitive, half defiant, twinkled with apparent love of fun. He soon devolved the out-door work on his assistant, although the office-work at the station was light, the drawings being prepared and the principal accounts kept at the engineer-in-chief's office in St. John's Wood. These gentlemen had arrived to superintend the works of the great line of Railway, for which contracts had been taken. Before long each of them had added a pupil to Mr. Stephenson's staff. The younger of these gentlemen lived to succeed that famous engineer as engineer-in-chief of the London and North Western Railway;[13] with the other the reader may become better acquainted before the close of these pages.

[12] George Watson Buck (1789-1854).

[13] This is William Baker (1817-78), who was articled to G. W. Buck.

CHAPTER II

Preparations for a change

In a country-town where almost everyone knows as much, or more, about his neighbours' business as about his own, the arrival and establishment of two civil engineers was accepted as an outward and visible sign that, for good or for evil, the great railway experiment was about actually to be made. Before long, one or two keen-visaged, weather-beaten Yorkshire and Northumberland men, neither graziers nor farmers, nor mechanics, but seeming to possess some resemblance to each of these active classes, found means of establishing themselves in roomy farm-houses, where the rickyard soon began to assume the appearance of a carpenter's yard. Then men set to work in a field very near the mail coach road, where the hill dropped rapidly to the valley, in constructing waggons of a form utterly inconceivable either to a farmer or to a carrier. Cast iron wheels, formed with the flange which is now so familiar to every loiterer at a railway station, but which was then an unexplained puzzle to most folk, were firmly keyed on to solid axles; a ponderous oak framework was fixed on the axles, and a special provision was introduced, for tilting up the large shallow tray which rested on that framework. A stout, hale, well-humoured Yorkshireman stood or sat in this field from early morning till sunset, and inspected the progress of the waggon with a never-flagging interest. People came to stare at him as much as at his labourers.

After a while the field in question, and a broad strip of land reaching away over the country north and south, were enclosed with a stout and ponderous frame of oaken posts and pine rails, a fence admired by agriculturists, and greatly misliked by sportsmen. Then long trains of brick-carts and timber-carriages commenced a service from the nearest canal wharf, and piles of bricks, and heaps of half-wrought freestone from Yorkshire quarries, began to accumulate near the waggon-building field. Then there was a great field-day of the engineers, a display of lines and levels and measuring tapes, an arbitrary infringement on the turnpike road, and the foundations of the great new viaduct were commenced. It was necessary to press this work as rapidly as possible, as some million and more cubic yards of embankment had to be carried over it and tipped to form the southern portion of the great Colne Valley Embankment. For seventeen feet the workmen had to sink, before arriving at a foundation sufficiently firm to sustain the weight of the piers. If that was how railways were made, people thought the work was simple enough.

Some two miles northward of this starting-point, the Northumbrian had commenced his operations. A tunnel of a mile long was a serious undertaking for the engineering practice of that day, and a costly, and as it proved, an unnecessary, method was adopted for ensuring the accuracy of the direction of the human moles.[14] The line through the tunnel was straight, and had been set out and pegged over the surface, as in other portions of the route.

At every furlong in length a shaft had been sunk, with the intention of opening a drift way (a small heading) from end to end, and thus running both line and level through under ground, before commencing the main excavation and lining of the tunnel, which it was intended to carry on through three of these shafts, properly enlarged, using the others for ventilation. So far nothing could be better. But concur-

[14]This is the Watford tunnel, which began a mile north of Watford station.

rently with this orderly arrangement of the miners came the over-careful science of the engineers. Three small observatories were erected, at equal distances on the centre line, and a solid brick pillar, detached from the wooden floor and staircase, which divided the interior of the observatory into stages, was built in each. A telegraph was attached to the roof, and a highly finished and accurate transit instrument, from the well-known hands of Troughton and Simms, was sent down to do duty in the wood pierced by the tunnel, and to check the lines dropped down the shafts to give the direction below. While the men of science rose in the air above, the practised northern miners bored below; the drift-way, carried on from eight piers at once, made more rapid progress than did the slow pedestal of the astronomic instrument; and the chief difficulty in ranging the line, arose from the draught that rushed through the heading when complete, the constant ascent and descent of inconsiderate workmen, or unwelcome visitors, and the knocking up of the engineers with acute rheumatism. Thus it came to pass that the elegant and costly transit instrument was used once and only once; that its cross hairs cut exactly on the lines that had been dropped down the shafts; and that, a week after the completion of the headway,—which week had been spent in knocking off elbows here, raising the roof there, and lowering the floor in another place, with occasional loss of time and of temper, as some intruder came blundering through the long narrow cavern,—the pupil of the sub-engineer, the only one of the staff left on his legs, had the extreme satisfaction of viewing the red signal lamp, fixed at the north end of the head-way, from the southern extremity, over a regular and exact line of candles, one close to each shaft. The question of the direction of the tunnel was thus solved, and the chief care now requisite was to ascertain the accurate range of the levels.

In 1833, works which are now regarded with comparative indifference, engaged the very anxious care of those who had the responsibility of designing them. Two or three skew arches with stone voussoirs, a road and a river viaduct, one or two road and occupation bridges per mile, and a mile of tunnel through the chalk, appeared to be an immense quantity of work, and the monthly measurement and estimate of the fencing, earthwork, and masonry, over twelve or fourteen miles, involved a great consumption of time.

Robert Stephenson, in those days, almost lived on the line, and the first occasion on which he visited the portion in question, after the contracts were let, accompanied by the Secretary and by four or five of the Directors, was the twelfth time that he had walked the whole distance from London to Birmingham. The personal appearance of that fortunate engineer is not unfamiliar to many of those,

14

whose eyes never rested on his energetic countenance, frank bearing, and falcon-like glance. It is rarely that a civilian has so free and almost martial an address; it is still more rare for such features to be seen in any man who has not inherited them from a line of gently-nurtured ancestors. In the earlier days of Robert Stephenson, he charmed all who came in contact with him. Kind and considerate to his subordinates, he was not without occasional outbursts of fierce northern passion, nor always superior to prejudice. He knew how to attach people to him: he knew also how to be a firm and persistent hater. During the whole construction of the London and Birmingham line, his anxiety was so great as to lead him to very frequent recourse to the fatal aid of calomel. At the same time his sacrifice of his own rest, and indeed of necessary care of his health, was such as would have soon destroyed a less originally fine constitution. He has been known to start on the outside of the mail, from London for Birmingham, without a great coat, and that on a cold night; and there can be little doubt that his early and lamented death was hastened by this ill-considered devotion to the service of his employers, and the establishment of his own fame.

The elder Stephenson had no share in the actual direction of the works of the London and Birmingham, and as his son found himself sit firmer in the saddle, he showed at times something of his father's determined and autocratic temper. He met his people with a frank and winning smile; his questioning was rapid, pointed and abrupt, and his eyes seemed to look through you, as you replied. Very jealous of anything like opposition or self-assertion, very unjust at times in suspecting such a disposition, he was disarmed by submission, and quieted by very plain speaking. On one occasion he had detected what he thought, and probably truly thought, to be a notorious instance of "scamping" on the part of one of his contractors. He sent orders to the man to meet him on the works, and coming up rapidly as he caught sight of him, burst out, "So-and-so, you are a most infernal scoundrel!" "Well, sir," meekly replied the delinquent, "I know I am." Robert Stephenson's lecture was at an end, and the man was dismissed with something like a friendly warning.

The lively interest with which a young man, on the threshold of his profession, watched every detail of the work of which he had practical superintendence, will not be shared by the reader after a lapse of two or three-and-thirty years. No great point of professional practice arose on this portion of our second English line of railway, unless it were the development and practical execution of the theory of the skew arch, for which, at that time, the engineer was referred to very elementary works. The conception of each arch-stone, or course of bricks, as forming portions of the thread of a large screw, 15

the pitch of which was determined by the angle at which the centre lines of road and railway intersected, and by the radius of the cylindrical portion of the arch, admits of very precise and simple determination. In these bridges (then new, as executed on so large a scale), models were made in deal of each voussoir, and when put together, and secured by a strap, the single line of arch stones stood perfectly firm in its miniature representation. The working of the stones themselves, costly as they were, at so long a distance of canal and common road transport from the quarry, was thus greatly simplified by the freedom with which any measurement could be taken from the model. The lines of the intervening courses of brickwork were marked on the sheeting of the centering by the aid of a flexible straight-edge, and the giant screws twisted themselves into place, without a check and without an error. Blackened by smoke they yet remain, as evidence of careful and sound workmanship, although in the present state of our experience the space would be spanned by iron instead of by masonry.

One point, however, of this early practice demands attention, because it is one that may often recur, and the dearly-purchased experience of 1835 should not be lost sight of. It has been mentioned that three of the shafts of the tunnel were intended for working shafts. At each of these, first a horse-gin, and afterwards a small steam-engine, was erected, for the purpose of raising and lowering the skips or buckets. The head-way, or small central tunnel, some four feet six in height, by three feet or three feet six in width, ran off from the bottom of the shaft, which was nine feet in diameter, and properly lined with timber framework. A difference of opinion arose between the resident engineers, as to the safest method of widening out the tunnel to its full dimensions immediately under the shaft. When the excavation was complete, and as the brickwork lining was keyed in, a cast-iron curb or frame had to be built into the top of the arch, on which, in its turn, the permanent brick lining of the shaft was to be supported. The method adopted in the first two cases was that suggested by the younger engineer, namely, to leave the shaft itself, and some six or eight feet on either side of it, with as little disturbance as possible; to sink it to the full depth; and then to run a narrow headway on the bottom level, and, at the distance of two or three lengths of centering (nine feet each) from the shaft, to open and complete some thirty or forty feet of tunnel in each direction. The ground being thus firmly supported, the miners worked back towards the shaft, and finally excavated and lined the shaft length itself, and fixed the iron curb in perfect safety.

In the third shaft, however, and after a change in the *personnel* of the engineering direction at this spot, a different system was adopted. It was determined to open the shaft length in the first instance, and to line it and fix the curb before opening out any further portion of the tunnel. A vein of sand showed itself through the chalk in the side of this shaft, but such faults were not uncommon in the tunnel. It unfortunately happened that during the operation the side of the shaft came in with a run, and nine unfortunate men were buried in the ruins. Besides this lamentable loss of life, the delay and evil occasioned by the accident were excessive, further danger was incurred, and the only resource left was to open the whole space of some thirty feet by forty feet square, up to the very surface; to build in the arch as if it had been a bridge, and to increase the masonry at this point to a large amount. The experience was dearly bought, and as the expense was borne not by the Contractor, but by the Company, it is evident that Mr. Stephenson did not consider it to be a case of ordinary casualty covered by the contract.

CHAPTER III

The first railway approach to London

The metropolitan terminus of the London and Birmingham Railway, according to their first Act of Parliament, was at Camden Town. A second Act was obtained, authorizing an extension to the site of the present station near Euston-square. Although no more than thirty-three years have elapsed, since this extension line was constructed, the transformation which has been effected in the locality is one that it is hard to realize. Since that date a new city has arisen on the site of fields and parks.

The Swiss Cottage, now a main starting-point for omnibuses and cabs, pending the completion of the railway that runs beneath the Finchley-road, then stood alone in wide and shaded fields, a real point of excursion for the Londoners who sought a few hours of country air. The Eyre Arms, a building then standing equally solitary, was occupied by the staff of Robert Stephenson, and was the headquarters of that gentleman as engineer-in-chief. Green, unbroken country lay between. The march of building had, in some places, crossed the Regent's Canal, and the attempts at villa residences on its bank, far from presenting their present dingy and depressing appearance, were not disagreeable places of abode. St. John's Church, or Chapel, however, marked the extremity of suburb proper, and one or two lines of untrodden road, covered with broken stone, were all the indications presented of a future extension of the dense mass of dwelling houses, towards the yet unbroken border of Sir Thomas Wilson's estate.[15] Unbroken, that is to say, by actual building, for the piercing of that property by three lines of railway, has now pushed the visible suburb closer to Hampstead than it was, in 1833, to the present termination of a continuous street. The line from Kilburn to Camden Town then ran through unbroken country; and not only so, but the advance of the Hampstead Road, considered as a street, had been so limited, that a thin crust of houses, as it were, only lined its course; and, with the exception of crossing Park Street and the Hampstead Road itself, hardly a single house of any respectable size was touched by the extension. To Park Street, the line ran south-

[15]See the maps of railways and estates in F. M. L. Thompson, *Hampstead,* 1974, 58-59, 102-103.

18

ward through fields of stiff clay pasture; from Park Street to Hampstead Road, its site was chiefly occupied by small and not very well tended market-gardens, and a little colony of firework makers had their cottages or rather huts in this intramural desert. South of the Hampstead Road, the fields and farm buildings of a great milk purveyor[16] reached nearly to Seymour Street. The legend of the spot, and of the day was, that the proprietor of this establishment had, for a long time, endeavoured to maintain on it a thousand cows, but that, do what he would, he could never exceed 999, and generally found himself some ten or twelve short of the round number. The animals occupied every field, but the main manufactory of milk was in the sheds and stables, where a diet of brewers' grains was thought to produce the valuable fluid in the largest possible quantity, although the quality was grievously inferior to that formed on a grass diet. Indeed the animals were thought to be so injured by food resembling, in its effects on their constitution, that of a constant habit of dram drinking among their human neighbours, as to be ultimately decanted or distilled into milk, and to yield a copious supply of unwholesome fluid till they were reduced to little more than skin and bone. It is for agriculturists to decide on the truth of this view; what is certain is, that a visit to the abode of the thousand cows was not calculated to give one an appetite for their milk. It is probable that the difference between the quantity of milk actually drawn from the cow, and that daily consumed in the metropolis, is now supplied by the aid of inorganic rather than of organic chemistry.

In the two miles of extension from Camden Town to Euston Square, the engineers had to solve nearly every problem which has subsequently to that time been encountered by the projectors of metropolitan railways. The canal had to be crossed under heavy penalties

[16]Thomas Rhodes, *ibid,* 218.

for interfering with its traffic. The alteration of an inch or two of level in the great highways was a matter of keen debate in committee, and the execution of the parliamentary conditions was closely watched by the courteous vigilance of Sir James Macadam. The sewers had to be avoided or provided for. Nearly half the bridges that were constructed were insisted on in order to provide for future roads, and intended streets and crescents. The gradients were of, what was at the time considered, unparalleled severity, so much so that the idea of running trains propelled by locomotives from the terminus was laid aside; a powerful winding engine was erected at Camden Town, and a cumbrous but well-considered apparatus of ropes and pullies was laid down, in order to draw the trains up the inclines of 1 in 75, and 1 in 66. The main difficulty which, at this stage of our scientific knowledge, was found to press on the engineers of the line, was the design and construction of a telegraph, which, in all states of the weather, and by night as well as by day, should instantly communicate the orders of the station-master at Euston, to the engine-men at Camden Town. Mr. Wheatstone was then but commencing the experiments, which have led to the perfect solution of a problem, impracticable to science without the aid of electricity.[17] The first efforts made in the study which has led to our present Atlantic cables, were directed to the transmission of musical tones through wires.

[17]For the introduction of the electric telegraph between Euston and Camden Town in 1837 see G. Hubbard, *Cooke and Wheatstone*, 1965, 48-55.

It may readily be supposed that the indignant astonishment, with which county owners and occupiers regarded the authorized trespass of the railway pioneers, was not greater than that evinced by metropolitan and suburban residents. Not rarely has it occurred for a respectable inhabitant of Mornington Crescent to find the peace of his family disturbed by the announcement, that strange men were clambering over his garden wall; and on sallying forth, indignant, to demand the reason of the intrusion, to find them coolly engaged, with hammers and cold chisels, in boring a hole through the wall of his tool-house or summer-house. One sufferer in particular discharged many torrents of fiery indignation on the devoted heads of Mr. Stephenson's staff. Public sympathy, it is to be feared, would even now rather be on the side of the residents than on that of the intruders. The gentleman in question was a schoolmaster, and eminent for that high respect for his own personal dignity which, from the days of Dionysius to those of Dr. Parr, all independent educators and chastisers of youth have been known to cherish. His school was not large, but it appeared to be respectable, and his house was rendered more attractive by the presence of a sprightly black-eyed daughter. It so chanced that this house occupied a corner of one of the great roads crossed by the railway; and that an annex of one storey, a hall or a breakfast room, with a flat-leaded roof, lay in the very centre line. To complete the survey on the large scale of 80 inches to the mile, which was the first duty of the Staff when the Act for the Euston Extension had received the royal assent, it was necessary repeatedly to invade this gentleman's premises, to cut holes in his garden-wall, and to make use of his very convenient "leads" as a station for the theodolite. For the sake of convenience and despatch, the surveying party always carried a short and handy ladder; and as the Act of Parliament authorized entrance on all the scheduled property for the purpose of surveying, the orders were, to take what is called "French leave," to proceed in the most straightforward and rapid manner, and to take no notice of occupiers unless they interfered with the survey. Whether this was the most courteous mode of procedure is not the question. It was legal, and it was a great saving of time. On a calm retrospect of the *pros* and *cons* of the question, it seems that courtesy for the most part would have been wasted, and time with it. However that may be, it could hardly have been very agreeable for the worthy tutor, while perhaps enforcing, with sonorous voice and uplifted finger, the due sequence of τετυφα, τετυφε, τετυφοιμι, to find the window of his school-room suddenly darkened, and, on looking up to ascertain the cause, to behold on the leaden flat, immediately outside the room and level with the floor on which stood his wondering scholars, a pair of very long legs descending from beneath a blue macintosh cape. The back of this figure was respectfully presented to the inmates of the house, as a sign that no intrusion was intended on their privacy, while the owner of the legs and one or two assistants were carefully fixing a theodolite on the flat, and marking with pen-knives the exact spot occupied by each of the three legs of the instrument, for the sake of exactitude and of despatch on future occasions. The first scene of this kind would probably have afforded the finest study for the observer of human nature, as, after repeated rehearsals, the eloquence of the man of learning was apt to partake too much of the vituperative to be strictly rhetorical,—regarding rhetoric, as Aristotle teaches us to do, as the art of persuasion.

The window is thrown open; the master, in flowing dressing-gown, advances majestically towards the intruders; the amazed school-boys, delighted with the "lark," crowd in a semicircle behind him; the half-scared, half-indignant mistress appears at an adjoining window; the pretty daughter makes good use of her fine eyes on the floor above (as is noted by the engineer's pupils); the giggling servants crowd on tip-toe in the yard below. "Who are you, sir, and how dare you come on my leads?" "Sir, I am an assistant of Mr. Robert Stephenson, and I am engaged on a survey for the Euston-grove extension of the London and Birmingham Railway." "And what   21 business have you on my premises?" "The centre line of the railway passes exactly beneath the plumb-bob of my instrument—" "Don't talk to me of plumb-bobs, sir; how dare you to climb up there, and to have the impertinence to stare in at my windows?" "If you will have the kindness to look at this book, sir—" "Don't talk to me of books; I say, how dare you come here?" "That is just what I produced the book to explain, sir—" "What on earth do you mean, sir?" "I mean, sir, that this is the Act of Parliament in virtue of which the officers of the company are authorized to enter on the properties scheduled in the book of reference for the purposes of survey." "Hang the officers of the company! and the purposes of survey! and you too, sir! Once for all, will you leave my premises immediately?" "I regret that my duty forbids me to do so, sir; but we will be as rapid in our work and give you as little annoyance as possible." "Then, sir, I shall instantly send for a policeman." "Perhaps that will be the most satisfactory course, sir. The policeman will, no doubt, convince you that we are only doing our duty.—Set that pole a little to the right." Grand tableau! Young people seem most highly to appreciate the fun. Doctor retreats to send for the police.

The fire-work colony met the invasion with less dignity, but with no less hostility. Among the occupants of the small market gardens was a man who bore the name of that great master of the English tongue, the author of the immortal "Pilgrim's Progress." The mantle of his namesake or ancestor had so far descended on this modern Bunyan, that he possessed a rare flow of vernacular English, and was, on any subject that deeply stirred him, a born orator. And he *was* deeply stirred. Leaning with one hand on his spade, and waving the other to enforce his periods, this man would harangue at once the offending Engineers and the sympathising audience, with much the same happily-worded flow of slightly inaccurate invective which would, no doubt, if heralded by the press, have secured him a seat for Rochdale, or for Salford, or for some great centre of reforming indignation.— "That the mean and sneaking railroaders should be afraid to venture near gentlemen's fine houses, and come with their peep-glasses and pick-axes to bore holes in poor men's cottages, and trample down the cabbages and lettuces in his poor little bit of garden, was a fine pass that Old England had come to. Pretty gentlemen, that you call yourselves, to come clambering over folks' walls, with your ladders and your hammers, and your levels, and your bevels, and your devils, taking the very roofs from our heads—why," concluded Mr. Bunyan, "look here! now I'm neither wind-tight nor water-tight!"

22 Another personage appeared at times on the scene, taking a lively

interest in any symptoms of a row, of whom it is marvellous that no satirist or moralist of the day ever caught the portrait. He would have been a fortune to a romance-writer, for he was an incredible, unimaginable, character; or rather, he was such a character as we have been wont to consider as the offspring of the most brilliant of human imaginations. He was the very double of Falstaff. In his size, his portly rotundity, his merry twinkling grey eye, his more than questionable jokes, his rollicking merriment, the sort of mingled awe and pity which he seemed to inspire in the lady, much his junior, who usually tended his steps, and who seemed divided between the apprehension of what he might say next, and the conviction that, after all, it could not be worse than what he had said before,— he was the very knight that Shakespeare drew. An old soldier, too, he seemed to be. Are there no records of the passage of so marvellous a dramatic character over the early scenes of the nineteenth century?

Step by step the feuds were appeased, or expired from want of fuel. The survey was finished. The sections were completed. Then for rapid weeks of indoor-work—designs, working drawings, specifications. Then a brief pause, when the completed engineering work was remitted to the awful recesses occupied by the Board of Directors; and then the announcement that Grissell and Peto[18] had contracted for the works, at a price amounting pretty closely to the value of a double line of sovereigns laid, touching one another, from Camden Town to Euston Square.

The magnitude and the scientific interest of the works of the Euston Extension, great as they were at the time of their execution, have been so far exceeded in importance by the more recent efforts of engineering skill, that the subject has ceased to possess any great

[18]Thomas Grissell and (Sir) Samuel Morton Peto (see note 62) were partners from 1830 to 1846 and became, with the Cubitts, the largest building and railway contractors in London.

23

interest, except as a question of the history, or rather of the chronicles, of construction. The chief work, as regarded novelty of structure, and also as combining some degree of architectural pretension with the mere requisites for strength, was the bridge over the Regent's Canal. The later and bolder arrangements adopted by Mr. Brunel at Chepstow and at Saltash, and by Mr. Stephenson at the Menai Straits, have taken the lustre off the method by which Mr. Fox, under Mr. Stephenson's approval, introduced a very great improvement, in this case, on the then existing plan, of constructing bridges of respectable span by a combination of parallel and transverse girders. The chief historic interest attaching to this bridge arises from the fact of a deplorable accident, which, some years later, fully justified the objections raised by mechanical men to the method of combining cast and wrought iron employed on some other portions of Mr. Stephenson's works.[19]

[19]This is the Dee Bridge accident. See p. 137.

A curious instance of the behaviour of concrete deserves, however, to be recorded. A considerable part of the Euston Extension was carried through cutting, protected by curved retaining walls of brick. The experience of the Primrose-hill tunnel, and the later experience of the walls in question, show that it would have been the best economy to throw an invert over the whole width of cutting, and to have constructed the walls of such a form as to throw the thrust,

to which their outer surfaces were exposed from the "creep" (a slow semi-fluid movement of the London clay), fairly on the line of resistance of the inverted arch. It has been found necessary to remedy this omission by the heavy and unsightly girders, which now strut the retaining walls from side to side, as a falling house is retained in the perpendicular by props or ties. In the original designs for the tunnels on the London and Birmingham Railway, all the sections contained inverts, of rather slighter depth than the arch. In the Watford tunnel, through solid chalk, the invert was found to be altogether unnecessary, and its place was supplied by ordinary footings to the side walls. A similar change, which was not only a saving of a large expense, but a means of very considerably increasing the rate of progress of the limiting works of the line, was accordingly attempted in the Primrose-hill tunnel; but the hard London clay, though requiring a pick-axe to cut it, soon showed such unmistakable menace of rising from the floor to fill the whole excavated area, that the original section was re-established. In the Extension this was unfortunately neglected, and the result is very palpable to the traveller.

The whole of the retaining walls were built on a bed of from three to four feet of Thames ballast concrete, the best available material for the purpose. To secure uniformity of gradient, an iron bench-mark was driven into every pilaster, and the level registered to the hundredth part of a foot. The gentleman who had charge of the operation found, on checking his book after the works were in full swing, that the levels were wrong. It was only by three or four hundredths of a foot that they varied, but extreme accuracy was desired. Some alarm was excited, and day after day the levels were revised, and always with discordant results. The instruments were carefully examined: no error could be detected in their adjustment, and moreover the constant use of even backsets and foresets—that is, of reading at an equal distance on each side of the observer,—which alone will neutralise any error in the instrument that is not of great magnitude, and easy of detection, had been scrupulously adhered to. The subject baffled all explanation for some weeks, after which it was found, that the differences in question arose from the swelling of the concrete after the brickwork had been carried up to the plinth of the wall, at which height the levels were taken; and that as the disturbance thus caused ceased, the bench marks all arrived at a perfectly true gradient, although the absolute height was some three-hundredths of a foot above that of the same spot where the level was taken. The experience may be valuable to those who have to deal with concrete under untried circumstances.

A very fortunate error led to an unusual, but by no means unnecessary provision for the traffic of 1866 in the year 1833. The construction   25 of works for four lines of rail, through property so expensive as that leading from Euston Grove, would have been thought by the Directors, and probably by the Engineers, of the earlier period, to involve an unwarrantable outlay of the Company's funds, had it not been from the expectation that the Great Western Railway would have joined the London and Birmingham at Wormwood Scrubs, and availed itself of the Euston Terminus.[20] The great difference of gauge, among other causes, led to the abandonment of this intention; but the result is that the London and North Western Railway is better adapted to the wants of its enormous traffic, than could otherwise have been the case. The only reason for regret on the subject is, that the same necessity was not thought to attach to the Primrose-Hill tunnel; which, however, was a part not of the extension, but of the original line.

[20]For this plan and its abandonment see E. T. MacDermot, *History of the Great Western Railway*, 1964 edn, i, 8–9, 19.

CHAPTER IV

Demand and supply of engineers

With the steady progress of the works of the London and Birmingham Railway, stimulated as it was by the unexpected results of the earlier experiment of the line from Manchester to Liverpool, and with the rising expectation of the shareholders, arose the first great impulse towards a more general system of railway communication, between the metropolis and the great centres of trade and of commerce. Bristol, Southampton, Dover, Yarmouth, as ports,—Greenwich and Brighton, as centres of coach traffic, were among the first applicants for parliamentary powers for this purpose. A fermentation rose high in 1835, almost like that which, ten years later, attained such extraordinary intensity. Here lay the great good fortune of Mr. Robert Stephenson, who possessed the signal advantage of a practical acquaintance, from childhood, with the details of mechanical work of all kinds, together with that of a regular education expressly intended to qualify him for the profession of a civil Engineer. He thus stood almost alone, and owed, no doubt, to this favourable conjunction of circumstances, even more than to his practical and his scientific acquirements, or to his knowledge of the world, the eminence which he so rapidly attained. In this respect he had the advantage of his great rival Brunel, who, possessed alike of hereditary constructive genius, of bold and courageous originality, and of the results of the most scientific training then attainable, had in some measure to make his acquaintance with the practical details of actual work, and with the best method of dealing with master workmen, at the cost of his supporters.

But railways were the cry of the hour, and Engineers were the want of the day. If they were not to be found ready made, they had to be extemporised. And so they accordingly were. Long before the school, in course of formation on the works actually in progress, could turn out men of sufficient age to appear in public as authorities on expenditure and construction, Engineers-in-chief were required by many a group of projectors. So came to the front,—military men, accustomed, perhaps, to rapid and skilful sketching of country, able with the theodolite and with the sextant, but unacquainted with   27 other requisites of their improvised profession;—cautious martinets, formed in the old school of the Royal Engineers, at the time when the Duke of Wellington, as Commander-in-Chief, found such difficulty in making use of these dependents on the collateral authority of the Master-general of the Ordnance, that he took the step of forming the Royal Staff Corps, to have Engineers of his own;—mining surveyors, accustomed to the use of the dial, and to the intricacies of subterranean work; mathematical engineers, good, no doubt, to direct the smithy and the lathe, but unaccustomed to works of magnitude;—architects who generally limited their claims to the construction of stations, and who seemed to be regarded as interlopers by all who, on any of the above grounds, called themselves engineers. So boldly, however, did these new commanders take up their positions in parliamentary warfare, that many of them established themselves in lucrative berths. Of one, who spoke with great authority before a Committee, counsel took the liberty to remark that, if he was so good an Engineer as he asserted himself to be, it must be owing to Divine inspiration, as he certainly had not had the ordinary human means of acquiring the necessary education. At a time when the profession of civil engineering is suffering from a cruel sort of ''lock out,''[21] it is most tantalising to reflect on the golden showers, that watered the early growth of the various grades of railway constructors.

The younger men, unable to front a public meeting or a board of directors, were in demand everywhere for field work. Engineers who had pupils to spare, *lent* them to one another, or *let them out* on terms of hire agreeable to all parties. Thus the scene of personal recollection may readily change from the busy hive of workmen, that filled the great open ditch of the Euston Extension, to the Derbyshire moors, the Essex corn-lands, or the Norfolk fens. The ordnance map was the great guide of the projector of these days; but few, even of those who called themselves engineers, seemed then to be fully aware of the immense advantage to be drawn from the careful study of this admirable chart. Surveys were made at large expense, which were, if not absolutely worthless, quite unnecessary; and in some instances tracings of the enlarged Ordnance plans were provided for the guidance of the surveyors and engineers, which, devoid as they generally were of writing, were far from being such useful and intelligible guides as the actual, published map. Engineers seemed to act like clergymen who are not gifted with rhetoric, and who yet seem to think it a duty to provide their hearers with a very inferior home-made article, rather than submit to the humiliation of purchasing an excellent one that happens to be ready made.

There was one instance in which more than a twelvemonth was occupied in surveying some twelve hundred square miles of country,

[21]The financial crisis of 1866 had led to the abandonment of many plans for railways previously sanctioned.

the whole of which was included in the published Ordnance map. A very beautiful sketch-survey was first made, and the fortunate draughtsman took rank at once, on the strength of it, as an Engineer-in-chief. He procured from the Ordnance authorities the calculated lengths of certain of their triangulated lines, and on these actually carried out a new triangulation of the country; fixing long poles on certain conspicuous points of the intended line, and including every spire and steeple within range of the observations, which were made by the box-sextant. A gigantic roll was prepared, on which the new points, or rather the old points now, for a second time, and by instruments of less precision, trigonometrically fixed, with the addition of the railway poles, were laid down, the idea being that a railway survey of inimitable accuracy could be thus produced. This was *after* the Act of Parliament had been obtained, on plans deposited on the usual four-inch scale. An enlarged survey as usual followed, but the plan of the line was *never laid down on,* or compared with, the trigonometrical sheet. It is true that occupation was given to a considerable staff, during a time that the completion of the railway was doubtful, by this introduction of the military element into the labours of the surveyor; but the shareholders might have questioned the advantage of the proceeding to themselves.

The idea of a grand railway route through the Eastern Counties of England to the port of Yarmouth, which, thus linked to the metropolis, was to distance even Liverpool in its fabulous increase of prosperity, was one of the schemes of 1835. A pupil from Mr. Stephenson's staff was among the subalterns whom the engineers of this undertaking impressed into the service of the intended company.[22] After receiving a decorous lecture, as to the exactitude requisite, both in the discharge of his scientific work and in the control of the expenditure he was about to incur, he set out on the exploration of a portion of this line, furnished with level, staff, chain, Ordnance tracings, written instructions, and a very moderate sum of money. Desirous to give full obedience to the excellent advice, which he had received with some awe, he engaged a place on the outside of the mail. With the advance into Essex the evening mist thickened into rain, and the journey was not half accomplished before the wet began to find its way through the "warranted water-proof" cap which formed part of the strictly limited wardrobe with which, as usual with most very inexperienced travellers, he thought it convenient to travel. It may be left to the military critic to decide how far the famous equipment of "a sabre, two towels, and a piece of soap," are fit to constitute the entire baggage of a commander-in-chief. For an engineer who wishes to preserve his own efficiency for a month or six weeks of field work, two complete out-door, and one indoor,

[22]A characteristic example of Conder's oblique methods of referring to himself.

29

suit, besides proper great coat, rug, and cape or dreadnought, are indispensable. The result of the economy and rapidity of motion aimed at in this instance was, that the young traveller arrived at the ancient cathedral city to which he was first directed to proceed,[23] about seven o'clock on an October morning, after a night sedulously calculated to unfit him for commencing his labours. The first duty, after a hasty refreshment of the inner and outer man, was to deliver a letter of introduction from the Engineer-in-chief, under whom he was acting, to an old friend resident in the city. He was received with very quiet kindness; and after delivering his credentials, and making some local inquiries, to which he received what seemed rather evasive answers, rose to take a hasty leave. "What?" said the host, "your breakfast is just setting in the next room." "But I have breakfasted at the inn." "What did you do that for? You cannot go without tasting some of our famous Norfolk sausages; and before you have done, a horse will be put to, and my young man will be ready to drive you wherever you want to go. What inn did you say? Tell me the number of your room there. I will send for your luggage directly. You must have no other head-quarters than my roof while you stay in the country." "My young man," the son of the hospitable merchant, not only accompanied the young Engineer through all the parliamentary campaign, but returned with him to London, and became a pupil of the Engineer of the line. Such was an instance of the hospitality of Norfolk, in the time of mail coaches.

The troubles of the start were not yet over. On approaching the coast, with the purpose of identifying the few unintelligible lines by which the tracing of the Ordnance survey (for the real map had not been forwarded with the instructions,) indicated the localities to be traversed by the parliamentary section, not a single line of

[23] Norwich.

30

boundary or division was discernible. A long, perfectly straight road ran from the verge of the fens proper, to the bridge separating that district from the sea-port terminus. The line was to run to this bridge, from a point some half a mile to the south of the other extremity of the straight road. The road in question *was not shown in the tracings*, and *the fens were under water*. One broad, unbroken sheet of inland lake was all that was presented to the eye. The orders of the Engineer-in-chief,[24] when referred to in this difficulty, were brief. He was one of those men who made a more than liberal use of expletives, now happily rarer than was the case thirty years ago. ''Set so-and-so to run his level along the road from Yarmouth to Acle, and —— the Fens.''

[24]John Braithwaite (1797-1870).

The comic incidents of these early parliamentary surveys were many, and were often irresistible in their fun. The real cause, no doubt, lay in the fact that so many men found themselves suddenly in positions which they had some obscure doubts of their perfect fitness to occupy. The want of ease had therefore to be made up by a certain fullness of self-assertion. Then again, while some men, not at once at home in their new duties, were daily learning better to discharge them; others, in every way less competent, were seeking to push the real workers from their seats. So very little *ad hoc* education went to make an Engineer, when Engineers were in demand, but some men, on the score of a moderate acquaintance with mathematics, or with chemistry, or gifted with nothing more than a facility of utterance, thought that they might at least come in as *Consulting Engineers.* Then the necessity of depositing plans and sections by the last day in November, gave a stimulus to their whole course of action which brought out the salient peculiarities of the actors in full relief. The stolid amazement of the country people, and the usual hostility of the farmers, rendered the actual instrumental work often the least difficult part of the duty of the day.

It is not, however, the wish of the writer of these pages to regard the early period of a great physical change in this country, and indeed in the civilized world, mainly as yielding the elements of comedy. The few actors in those busy scenes who may yet remain, will not be discontented to have their recollections called back to a time very worthy of the pen of a chronicler, but which contemporary literature was not led to seize and to dwell on. As actors we were too busy to write, except reports, estimates, and accounts. The Engineer has of late, indeed, made his appearance in works of fiction. But it is the Engineer of the novel—not of real life. The truth was at once better sense and better fun than the fiction. That education, by which the public taste was led first to dwell on Dutch pictures of humble, and indeed vulgar, life, to retail the conversations of nurses, grooms,

31

coachmen, and the like, had only just commenced in 1833. Had the present magazine literature been then in existence, or had sensation writing been at that time as far developed into a commercial art as is now the case, very much, at which we may be content to hint, would no doubt have been fully described and highly flavoured for a public that desired only to be amused. But if picturesque detail, or good-humoured gossip, or appeal in some form to the appetite for amusement, must be necessary accomplishments of the writer who seeks to interest any but a very limited circle of readers, there may yet be a higher aim than the production of laughter. We stand at this moment, physically and mechanically speaking, on the threshold of a new world. We are pausing after the accomplishment of the first phase of the greatest revolution which has ever been effected in man's relation to the world in which he dwells. We have given him a speed of transit which he could not have attained if he had, instead of the services of the steam-engine, acquired the actual endowment of wings. We see a great check interposed in our course; the men long active in the service of their fellows, as regards the development of the means of intercourse, are now folding their arms in enforced inactivity. The six hundred millions which we have spent on our own railways, although, on the average, yielding a fair return on the par price, have been lavishly and inconsiderately spent. If the fourth part of that sum, which we may reckon, without exaggeration, as having been made into ducks and drakes, were now forthcoming for the necessary development of the feeders and ramifications of the great trunk and branch lines, an immense impulse to our national prosperity would ensue. There may be wisdom, then, as well as relaxation, in looking back to inquire how the chief industry of the last thirty years came to start *in a wrong groove;*—how it was that the service of the public was injured for the benefit of private individuals, and that a false direction was given to an industrial development of such unparalleled importance. The most sanguine ideas, of the most sanguine speculators, never contemplated the enormous traffic developed and created by the railway system. This gigantic and unexpected excess over the estimated traffic has been claimed by the projectors of railways as a set-off against the enormous excess over the estimates of their construction. The balance has been fortunate, and of course, to some extent, unexpected traffic has caused unexpected outlay. But those familiar with the subject know that comparatively little of the actual waste is thus to be justified. That our present traffic might have been conducted as conveniently, and more conveniently, than is actually the case, while at the same time it should have been weighted with some ten millions sterling per annum less of interest to be paid out of profits, probably few persons qualified to form an opinion will doubt. It may therefore be a service to the conductors of the public works of the future, to open an un-recorded chapter of the history of those of the present, and to point how from small beginnings, and from natural and neglected causes, arose part of that present state of suspicion and of hesitation which is paralysing so much of the energy of the world. If the attention of the country were still called to the steady progress of duly considered remunerative public works, carried on by intelligent enterprise, and not by blind gambling,[25] we should find the public mind less nervously discouraged by the armature of a fortress on the Alps, or by the sudden illness of a Turkish Vizier.

[25]Conder was writing while the effects of the crisis of 1866, which had arisen in part from "blind gambling", were still working themselves out.

That new men, called to positions for which they had never been prepared, will err, is a very evident proposition. That there existed in the country a mass of practical science, by a careful use of which *technical error* might have been so far circumscribed as to be reduced within small limits, those whose memory goes back to the time in question are aware. The evil to the country arose from the fact, that the best men were not made the best use of. The unavoidable evil was aggravated by much that was pure surplusage. Had the capital of railways been spent by men taking an enlightened interest in its application, and controlling that application with the simple aim of profitable investment, our position at the present moment would have been something very far different from the actual fact.

It has been a happy thing for the credit of the Engineers of Great Britain that they were, as a rule, paid by salaries, and not by commission. Had the latter mode of payment, which in many instances has much to recommend it, been sanctioned or insisted on by Mr. Stephenson and his earliest colleagues and pupils, no integrity of private character would have been enough respected to avoid the reproach of outlay without other motives than that of earning large commissions. Those who remember the early anxieties of that eminent man—professional anxiety only is intended,—would be well aware how unfounded such an assertion would have been in his, and in many other, cases. In some, however, such a suspicion would not have been unfounded. It is the desire of the writer to say nothing that can cause undeserved annoyance to any one. For those cases which may be referred to as beacons rather than as examples, instances will be selected in which the actors are far removed beyond the verdict of any human tribunal. To select none but meritorious men for portraiture would be an attempt to falsify human nature. It is rather from a contemplation of our past failures, than from that of our partial successes, that we must hope to derive lessons for future improvement.

It will no doubt be in accordance with the sound judgment of   33 all professional men to say, that the success or failure of their brethren has mainly depended on their professional industry. Take one man who was quiet and plodding, raised, perhaps, far beyond what his education or his acquirements gave him a right to expect, but steadily devoting his time to strict professional labour, and that man, as a rule, became eminent, or at least respectable and respected. Take another man of more varied talents and more showy acquirements, who turned his energies into another direction; who was brilliant under parliamentary examination, quick in throwing together a prospectus, successful in influencing directors, or in commanding the confidence of the Stock Exchange; and if that man did not himself get into trouble, his shareholders were pretty sure to do so. The fable of the hare and the tortoise had many illustrations in 1835 and in 1845. We may say so without fear of hurting the feelings of the hares, for most of the poor animals are dead.

Indeed the havoc that death has made in the ranks of a profession, which might expect to be distinguished by unusual longevity, has been most remarkable. Brunel, in the judgment of those who remember the iron energy of his youth, should now be a man in the prime of intellectual vigour. Robert Stephenson might naturally have looked forward to many more years of quiet authority. Locke, Rendel, Moorsom,—how many are the names which a greater reticence of labour and more attention to the requirements of health, might have kept for many years from the obituary![26] In regarding such a mortality it is difficult not to search for some cause peculiar to the profession. One sufficient cause may perhaps be detected in the habitual loss of the usual repose of the Sunday. For men to turn night into day is in itself a hard strain. Twelve days' work per week will try the strongest constitution; but make the twelve into fourteen and the fatal result arrives with startling rapidity. And working by day, and travelling by night, make a constant and unrepaid demand on the vital energy of the brain. The cost of the English railways includes the lives of many eminent men.

[26]None of the five engineers mentioned here attained the age of 60.

CHAPTER V

Work in 1835.
Preparation for Parliament

There were good times for Civil Engineers in 1835. Most persons who could lay any claim to the designation found their account in so doing in that year. The form in which the traffic of the future would be conducted was then becoming apparent, from the experience obtained on the Liverpool and Manchester Railway; and the great centres of commerce became anxious not to be distanced in the impending race. Bristol, Southampton, Norwich, Yarmouth, Dover, and Brighton, demanded to be put in communication with the metropolis, no less loudly than Birmingham, Liverpool, and Manchester. As competent Engineers-in-chief were at a premium, so also were all the elements of an efficient engineering staff. Young men who could handle the level and the land-chain, suddenly found themselves persons of importance. At no previous time, at few subsequent periods, before the great mania of 1845, was there so much pressure for work, such good pay, and such good fun. Engineering, as carried on in those days, involved both hard work and good fellowship. As the period for depositing the plans required by the standing orders approached, the offices of the busily occupied leaders of the profession became scenes of toil or of scramble. Night was economised as the days grew short. An impulse was given to posting; and chaises and pair, and chaises and four, were to be seen at not unfrequent intervals galloping east and west and north, at a speed rarely to be witnessed, apart from the impulse of the Engineer, except upon the road leading to Gretna.

As the fatal 30th of November, the day by the midnight close of which the plans must be lodged with the clerks of the peace of the several counties traversed by the projected lines, (on penalty of being thrown out of Parliament by the standing orders' committee,) drew closer, the excitement became more intense. A line which, in the magnitude of its misfortunes, stands unrivalled among the original trunk railways,[27] was brought into Parliament under the joint direction of two very eminent men; one of these, who has been more distinguished for some of his noble foreign works than for his practice

[27]This is the Eastern Counties Railway.

[28]Charles Blacker Vignoles (1793-1875). His greatest work was the road bridge over the Dnieper at Kiev.

[29]Braithwaite.

in England,[28] was gradually elbowed into the position of consulting Engineer. He was chiefly known and dreaded by the staff, when busily occupied in the large premises of his colleague, by his method of evincing displeasure when anything went amiss. It was a method often witnessed on the banks of the Seine, but which does not tend to get the greatest amount of work out of the educated Englishman. It consisted in gesticulating in the midst of the room, swearing in a loud scream, and with singular originality and profusion of blasphemy, enforcing the volley of startling oaths by stamping on the floor, and finally plucking frantically at the hair. The *ultima ratio* of the angry Frenchman, trampling upon his own coat, was the only point omitted. Somehow or other this authority swore himself out of power. His colleague[29] was a man of another order, except in the matter of swearing, in which it would have been difficult justly to divide the palm between them. A hearty, jovial man,—one of those who are said to be kind-hearted, because they are singularly frank, and enemies to no one but themselves,—the latter Engineer was popular among his staff, and numbered many attached friends. He had, in early life, run through his patrimony; he had married, and, it was said, run through that of his wife. With admirable gravity and equanimity he was now preparing to run through as much of that of his shareholders as they would allow him to spend for them. In this he proved eminently successful, and a line which, from the nature of the country which it traversed, ought to have been as remarkable for its low cost of construction and of working, as for its large and reliable traffic, grew year after year to become a by-word for unsafe travelling and for invisible dividends.

''Now, boys,'' cried the chief, coming one day into the thick of the work, and springing at a bound on to the mantelpiece, ''burn your night-caps, for the *angel* a one of them you'll see till after the 30th. Then you shall have a week to lie in bed.'' Accordingly for some ten days the labour of plotting sections, copying plans, numbering and copying references, and the like, went on almost without intermission. At nine in the evening would appear mighty bowls of oysters, gallons of ale, and other materials of a rude but hearty repast. A respite of some three-quarters of an hour would be filled up by uproarious hilarity, and then a fierce objurgation from the chief, the moment before the chief reveller, for so scandalous a manner of wasting the company's time, would set all briskly to work again.

It is true that the quality of the work thus performed was not altogether equal to that which a more sluggish rate of proceeding might turn out. If you prick through a dozen sheets of drawing-paper at once, a very slight deviation from the perpendicular in the needle is enough to make the twelfth plan very different from the first. Inaccuracies of this kind might be very harmless while you had to do with parish clerks or county surveyors; but the great feature in the parliamentary fight was the opposition. Recalcitrant landowners, hostile corporations, or even rival companies, set their own engineers to work. With more time at command for criticism than for construction, these opponents would sometimes go to the expense of taking tracings of all the deposited plans, and then denouncing to the expectant legislators their want of identity. One Engineer, on being thus taxed, set down each difference of detail to the unequal expansion and contraction of paper under different conditions; but this admirable practicable defence did not meet the success to which it was entitled by its originality.

The crisis came on a Monday. The farthest distance that could be traversed in a given time, by the best-paid post-boy, had been carefully studied. In the outlying counties the deposits had been sent off by the Saturday. On through Sunday and Sunday night toiled the diminished staff. The last post-chaise was in waiting at the door. The last minute at which it must be off to reach, let us say Romford, five minutes before midnight, was at hand. Without his coat, the engineer-in-chief was with his own hands completing the last book of plans, by the unsophisticated process of plunging his hands into a large saucepan of paste, and spreading the cohesive substance with his fingers; while one or two assistants turned over the edges of the sheets. Superfluous paste was removed by a rapid action, passing each hand between the convenient shirt-sleeve of the other arm and the side of the waistcoat. At last the final sheet is pasted in, the book closed, the expectant messenger tumbled into the chaise with all his credentials; whirling off with a shout, and with the proper accompaniment of an old shoe flung after the carriage for luck; and now, if wheels and horseflesh hold, and no sleepy turnpike man make undue delay, the deposit for the standing orders is safe. The wearied may repose, the strain is taken off, and men begin to fear that they will hardly be able to sleep if they go to bed in a regular way.

When those offices that had to make distant deposits had to close during the November daylight, there were others that prolonged yet an hour or two more the struggle against time. A suburban line, which now daily conveys an enormous stream of traffic over its unsightly arches, was in the hands of an Engineer who had gained his experience, such as it was, in the military branch of the profession. The work was carried on in the comfortably appointed offices of a surveyor, who afterwards rose to be, for a brief space, one of the busiest and most noted of contractors.[30] Abundant light was shed

[30]The line is the London and Greenwich, the engineer Col. George Landmann (1779-1854), the contractor Hugh McIntosh.

37

on the scene of toil by gas and by wax candles. On a large plate-glass drawing-board the final section was being traced at one end, while the Engineer-in-chief, with a clumsy old pair of brass dividers, and a steel straight-edge, was attempting to lay down the balance line on the other. As the run of the country from London to Birmingham had enabled Mr. Stephenson to adopt a maximum gradient of sixteen feet in a mile, or 1 in 330, this inclination had come to be regarded with the deference due to a law of nature. Hence arose sore perplexity to the worthy Colonel of Engineers. The substantial black pencil lines that he scored one after another on his section seemed all equally to disgust him. "I cannot tell how it is," he exclaimed, in bitterness of spirit. "When I reduce the embankments I get such tremendous cuttings, and when I diminish my cuttings I run into such a devil of an embankment!" He spoke with justice, if not with propriety; as his attempt to fit the gradient applicable to one district, to the actual section of another, was balancing him between excavations of 300 feet in depth, and embankments of 100 feet high. How the steel straight-edge and the glass tracing-board finally came to terms, in time for a deposit of the section, cannot be told here, for sleepiness became more irresistible than curiosity.

With the deposits of 30th November came rest, but not termination of employment. Still was there much to be done in the way of check and revision; and when a line was actually in committee, the stress was at times very severe, and a surveyor had to be sent down, as fast as horses could carry him, to verify some contested statement. Then, too, it might not be with the elements of time and of distance alone that he had to contend. Grim human opposition was raised by sturdy occupants, wroth to see their fine crops of "turmuts" decimated by the land-chain; or by yet more substantial owners, eager in their hostility to a scheme that would triple the value of their estates, and pay them, for a small slice, enough to disencumber the remainder. To whirl out of town at nightfall, fresh from the heated atmosphere of the committee-room, falling asleep as the rattling post-chaise emerged from the outward zone of the gas-lit streets into the cool calm of the country; to wake up, after the nightmare disquiet of a broken sleep, under the golden hues of sunrise, amid the first leaves of spring, still speeding on at a rate which, if less rapid, was so much more enjoyable than that to which our iron roads have accustomed us, was a pleasure such as rarely occurs to the busy man of the present day. On arriving at the scene of the new and urgent professional task, it might be that the foe was on the watch, and all sorts of military manoeuvres had to be practised in order to secure the object of the excursion. On one occasion, an 38 Engineer who was running a line through Norfolk, was warned of

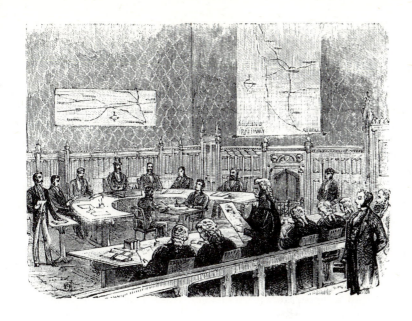

the strenuous and decided opposition which a certain landowner, a famous and unscrupulous parson, who was also a county magistrate, had threatened to all railway pioneers. Measures were taken accordingly. The wearied youth seemed to himself hardly to have fallen asleep, before he became dimly conscious that the moon was tapping at the window of the bedroom. The tapping was persistent, but it became evident after a time that the agent was not the planet, which was high and bright in heaven at the time, but the Boots of the inn; and to the remorseless knocking were soon added the unwelcome words, ''Half-past four, sir; you was to be called at half-past four!'' The two or three men, who are necessary for even the most economical mode of running a section, had, of course, to be duly warned the night before. Evil fortune arose from this necessity, for on arriving just before the point of day at the forbidden boundary, the surveying party found a garrison on the alert. A stout man in a round hat and leather leggings was leaning on a spade behind a hedge, and received the invaders with the threat, ''Mind what you are at. I am a county magistrate. I warn you off my land. Come on at your peril!''

When an Act of Parliament has been obtained, the entrance has been often practised in spite of opposition. When the Act is as yet in question, a defendant such as the one last described has the game more in his own hands. A parley was called, the urgency of the case was represented, and the invariable desire of that invisible bene- 39 factress of all neighbours of the projected line, "the Company," to compensate for all and every inconvenience, was dwelt on in every form. The Engineer's eloquence was fruitless. The parson was as obstinate as if the controversy had been on a theological point. At last the other was induced to ask the militant churchman to point out the limits of his property. "I am not bound to do that; find it out yourself, and enter it at your peril!" "Of course you are not bound," was the reply; "but there is such a thing as courtesy. I have a duty to perform in this part of the world, which I shall do sooner or later. You are resolved to keep me off your ground. We may be watching one another for a fortnight. On the other hand, if I know exactly the land you are so anxious about, it is possible that I can do my work without entering it. In that case I could give you my promise, and save us both much trouble." The parson was mollified. "Come," said he, "I cannot let you enter, for I have pledged my word to the opponents of the Company. But my boundary lies there, and there; and if you promise not to pass it, I will give you no further inconvenience." Happily, there was a footpath through the churchyard, which enabled the Engineer to turn the little outlying corner of the reverend gentleman's property; and less time was consumed in getting round the obstacle than had been devoted to the conversation.

CHAPTER VI

The duties of the Sub-Assistant Engineer

The Sub-Engineer, in the United Kingdom, is usually placed in a position at once of higher responsibility, and of harder service, than is the officer of the same grade in France or Italy. By the want of organization, which is characteristic of the English people, additional stress is thrown upon individual energy. The supervision of professional labour is rudely and imperfectly attempted among us; and the result is neither greater excellence of work, nor greater ease in its performance.

For an Engineer of the third class in the *Ponts et Chaussées* of France, or the *Ponti e Strade* of Italy, the course of operations is very distinctly traced. Under the general directions of the Engineer-in-Chief, an Engineer of the second class, answering to our Resident Engineer, minutely divides the several duties of his subalterns, receives their reports, and inspects their proceedings. The absolute work, however, of section or of survey, is carried on, not as with us, by the hands of the Sub-Engineer, but by a body of men under his orders, to whom we have nothing that corresponds in Civil Engineering in this country. Able contractors' foremen, or the private sappers and miners of the Royal Engineers, represent with us a grade of civil privates, so to speak, who discharge on the continent the duties which our young Engineers, perhaps only just out of their articles, are expected to perform, in addition to those of supervision of work, and of the preparation of detailed drawings.

Thus in the early times, when railways were first brought before Parliament, the duty that fell on the Sub-Engineer was often very severe. The politics of the line, so to speak, occupied the Engineer-in-Chief far more anxiously than did the actual engineering questions. The filling up of the share-list, the management of the directory, and perhaps the checkmating of the secretary, made imperative demands upon the time of the Engineer of a projected line. Resident Engineers, as a rule, were only appointed when works were about to commence. It followed that young men were frequently sent to learn their business by running sections and surveys over a country, with no other guidance than that of a line hastily drawn on a sheet of the Ordnance Survey.

Although the Ordnance Survey was, in perhaps every instance, the first basis of operations, it is remarkable how little use was made of the full and accurate information which is to be drawn from that excellent map. A careful study of the features of the ground may in many cases be made, with more advantage, from the shaded map, than from a cursory view of the localities; and in several instances much time and money was consumed in making a general plan or sketch of a country, which, when complete, was every way inferior to the engraved sheets that were to be purchased for a few shillings. It is possible that it is only the careful examination of country with the map in hand which can enlighten the Engineer as to the full merit and utility of the Ordnance Survey. The enlarged Irish survey, indeed, has been found to supersede almost entirely the necessity for original work in surveying, but much unnecessary field-work has been carried on in England.

In a line to which the writer has occasion more than once to refer, as the scene of some of his own earliest experiences, the instructions given to the Sub-Engineers, who were sent out to run the Parliamentary sections, were based on tracings of the enlarged Ordnance plans, showing fences and divisions of property. The line of the proposed railway was laid down on so narrow a strip of survey, that identification of the spot was extremely difficult, and the attempt at accuracy only induced confusion. A clear red line drawn on the engraved map would have been far more valuable as a guide on the ground, particularly as the survey from which the tracings were made was so old, or the tracings were so incorrect, that many of the fences were indistinguishable; and one of the most marked features of the district, a new turnpike-road of nine miles in length, which ran in an unbroken straight line across the fens, was not marked on the tracings!

A young man not yet out of his articles as an Engineering pupil, arriving alone in a strange country, furnished with such inadequate plans, and expected to send up a section of some twenty miles of country in a limited time, naturally felt somewhat taken aback. Staff-holders, chain-men, and other necessary attendants, had to be sought, to be hired, and to be instructed,—a labour in addition to that necessary for the conduct of a survey, which must be incurred in order to be appreciated. Then it was necessary to devote a certain number of days to the exploration of the country, and identification of the line—a labour which may be combined with that of levelling the actual section by a judicious use of the Ordnance map. Arriving at nightfall at a provincial town in the centre of the allotted district,

a wide sheet of water alone met the eye. On the following morning the scene did not appear to brighten. The small indications of locality afforded by the tracings were altogether obliterated. The road which perforce was to be the basis of operations had been ignored in the instructions. The wind rose in fitful gusts, shaking the instrument so as to make observation almost impossible; but not having the counterbalancing advantage of keeping within doors the swarms of little boys who drummed on the legs of the theodolite, or pressed down the wrong end of the telescope. It was a very dreary and discomforting commencement of a great national work that was made under these circumstances. If it was thought ominous of swamping the shareholders, the omen was not unfulfilled.

It was necessary, however, to run an actual section over the intended site of the railway, after leaving the diverging turnpike-road. The line was not inviting. When the dead level of the fens was quitted, swamps and morass were interspersed with arable land, and village greens or gardens.[31] The line ran through the midst of a decoy, where the swamp was so elastic, that the mere act of bending the head from the eye-piece of the level to observe the state of the spirit-bubble, threw the line of collimation above the top, or below the shoe, of the staff. It was only possible to obtain accurate levels across this morass by spreading the legs of the instrument to their widest possible extent, and kneeling between them to observe,—a pleasant occupation for a damp autumn day.

Crossing a deep navigable river, and thus departing from the convenient proximity of the mail road, the line ran, beyond the decoy, along the side of a series of gently undulating hills,and was freed from any further engineering difficulties, except a swing bridge of some magnitude. The section had scarcely been plotted inland as far as the provincial capital, before the Engineer-in-Chief made his appearance on the spot. Not, that is to say, on the site of the line, but at a comfortable dinner-table in the city.

This gentleman was accompanied by another, whose name became subsequently very well known in the railway world. He was, in his own opinion, an eminently scientific man. His mathematical attainments, whatever they were, were not called into requisition on this occasion. His eminence as a chemical discoverer, resting on the same personal testimony (namely, his own) as his mathematical acquirements, was then seriously controverted, by those who declared that the assertion of the commencement of an important experiment by boiling mercury in a tin vessel, was a fact that sufficed to relieve the reader from the trouble of further investigation of the reported process. As to Civil Engineering, it was patent that the good man's mind was in an altogether virgin state. Yet, appearing in the first instance as a director of the railway—a function which financial reasons

[31]The section of the Eastern Counties Railway on which Conder was employed was that between Norwich and Yarmouth. Much of the country was still undrained and very wet in winter, although none of it formed part of ''the fens''.

threatened to terminate—the endeavour of this notable person was to procure an appointment as a species of Consulting Engineer, or scientific Director, of the line. As it became more and more evident that the valuable character of the services he could render was not appreciated by the managers and officers of the undertaking, his tone and temper, with reference to the scheme, underwent a total and violent change. A personal hostility sprang up with the secretary, and the literary amenities exchanged by these two angry writers are among the curiosities of journalism.[32]

[32]The scientific gentleman was John Herapath (1790-1868), a pugnacious controversialist, founder of the *Railway Magazine* in 1836, which became his celebrated *Railway Journal.* The secretary of the company was J. C. Robertson. Their controversy is reflected in the *Mechanic's Magazine* (which Robertson edited), 1837, xxvii, 71.

The status, or scheme, or plan, of the scientific Director, had not been altogether ascertained or matured, when he came down to the fen counties as the self-appointed *adlatus* of the Engineer-in-Chief; who, perfectly frank and jovial in his own manner, was wont to cast queer, sidelong glances at his neighbour from time to time.

To point out the line over which the section had been taken, was the duty of the Sub, and it was performed from the driving-box of a post-chaise, in the interior of which vehicle the Engineer-in-Chief and the Director were ensconced. The day was cold and inclement, and the proposal to walk over the line was peremptorily declined. So was that to take the nearest route, diverging down bye-roads to the points of crossing. The chaise rattled along the mail-road, the various marked points on the opposite side of the river being indicated as they came in sight.

It soon became evident to the Sub that his chief, whether from an ancient and unforgotten knowledge of the country, or from the stores of a very fertile imagination, was giving to his companion such accurate details of the country denoted by the section, that the occasional stoppages of the vehicle, in order to point out any features of interest, seemed somewhat to interfere with the flow of description in which he indulged. Contenting himself, therefore, with the offer to reply to any further questions that were put, the young man, to whom a gentleman of the county had given the valuable assistance of his son as a companion, arrived very cold and very hungry at the Port,[33] where the scientific gentleman lost no time in cross-examining the waiter of the chief hotel as to the contents of the larder and the cellar.

[33]Yarmouth.

The investigation, long and minute, was somewhat of a waste of time, for it concluded with "Bring up all that you have. Coffee and tea, and some sherry, and some good cognac, and a bottle of champagne, cold meat and cold pies, and a chop or a steak as soon as it can be cooked; and make haste." A very hearty breakfast was well earned by the drive, and local friends, dropping in, so consumed the remainder of the day, that it became too late to return to dinner.

44 Then our scientific friend showed himself equally provident in the

matter of tea—the magnitude of the late breakfast having rendered dinner superfluous. He gave minute and precise instructions as to the preparation of the meal; emptying the canister into the tea-pot, and supplementing the hot toast, about which he was very particular, with a broiled fowl and mushrooms. "Waiter—the bill!" he exclaimed in a lordly tone, when the horses were put to. It was a long slip of paper. He examined it cursorily and with a magnificent air. "It is all right. B—, you had better pay it," quoth he; "I have no change. It all comes out of the same pocket, you know." The Engineer gave a very meaning glance, and paid the bill.

On another occasion the scientific gentleman distinguished himself in a more memorable manner. The sections had been long deposited, and the Bill was in Parliament, when it was thought fit that a deputation of four provisional Directors should make a sort of progress through the agricultural district traversed by the line, in order, as far as possible, to disarm local opposition. The same Sub-Engineer had been sent into Essex, to check the levels of a certain district, in which it had been discovered, in the course of the fight before the Committee, that a certain river crossed by the railway passed *up-hill* to a height of more than ten feet. These anomalies *did*, every now and then, appear in the more rapidly constructed sections. The Engineer-in-Chief directed his Sub to attend on the Directors, if they wished, when they came into the district which he was examining.

One fine day, therefore, about noon, the young man reluctantly broke off work, and marched back to the county town—Colchester, let us call it, or Chelmsford. Level and staves were glittering on the shoulders of his men, when a dust and a rumble came along the highway. A large yellow carriage, drawn by four post-horses, came thundering along, and an arm was waved out of the window. "We thought you were coming," said the scientific gentleman, when the other waited on the deputation at their hotel; "we saw your tools on the road." The "tools" were the level, staves, and chain, in the language of science, at least of this gentleman's science.

"Mr. B— desired me to offer any services that the Directors might require, and I have come here in consequence."

"Yes, yes—quite right. As he could not come with us himself, we desired him to send you. Now, there is no time to be lost. We must be at the Town Hall in less than an hour. I was just preparing my speech. See these blue books! Now, just look through them, and select every passage that you think will be applicable for my speech. Turn down the leaves, and score them in pencil, and then write a short abstract of each quotation, and then put down any remarks that strike you to connect the quotations. You have not a minute to lose, for I must go to luncheon before the public meeting. 45

I can't afford to trifle with my health."

The young Sub, of whose hunger no such account was taken, had not made any very great progress in this novel part of field-work—the preparation of a speech for a scientific gentleman—when the four Directors, prepared for their duties as far as the commissariat department was concerned, came hurriedly into the room. "Now, Mr. C—, where is my speech?—it is time that we were in the Town Hall. You will be so good as to keep close to me. If my memory should fail me for a moment, just whisper in my ear what I ought to have remembered. I shall call upon you to make a statement to the meeting, and you will see from my way of putting any questions to you what you have to reply. We all put the highest confidence in you."

It was not fated for all to go off in so simple and satisfactory a manner. When the Directors stepped on to the bench in the County Hall, an awkward silence succeeded. The room was half or three parts full, but every one looked his neighbour in the face, and no one seemed to have anything to say for himself. The Directors looked sheepish and confused. Something was wanting—no one knew what.

At length a tall, slender, and very grim Quaker arose. His distinct, harsh voice sounded through the hall with unmistakeable emphasis. "I should like," quoth he, "to know why we are called together, and who has convened this meeting." An awful silence followed. The Quaker repeated his remark a second and a third time, with the same result. "Because," he continued, in a yet harsher tone, "I believe that there are certain penalties provided in the case of persons calling together an assembly, and having no means of regulating or conducting the proceedings. I came here in consequence of an advertisement in the county paper; and, once more, I ask who is responsible for this meeting?"

The scientific gentleman opened his mouth and gave two or three gulps. That was all his speech. Then he collapsed, looking pale and extremely frightened. Then two of his coadjutors poked up the youngest of the four, who, with extreme confusion and much stuttering, expressed regret that the county member, who was expected of right to take the chair, had not honoured the meeting by his presence. Then the Quaker proposed, in order that no more time should be lost, that the last speaker, whom he regarded as the convener of the meeting, should take the chair, and proceed to business. This produced the counter-proposal that the Quaker should take the chair. He declined, but finally yielded to the general request. "I cannot say with truth," said he, "that I am grateful for the honour of being called to take this chair, for I consider it a great dishonour and great disgrace, both to this meeting and to myself. In the very awkward state of things, however, which has been brought about, it seems the only course to take; and I therefore call our friend So-and-so to explain the views of those whom he represents.''

The spirit of speechless alarm, which appeared to have seized upon the unlucky deputation of Directors, was only partially laid by the constitution of the meeting. One after another essayed to speak, but the result of their efforts was not reassuring. One of them, bearing a name known both in the City and in the House of Commons, addressed himself to the question that railways were not joint-stock enterprises.[34] If such had been the case he (the City Director) should not be there to support one. People had objections to joint-stock enterprises; so had he: there had been lamentable failures among them. But as the railway which he had the honour to support was *not* a joint-stock enterprise, he thought that objection would not apply; and therefore he trusted that the numerous and influential meeting which he had the honour to address would support the railway of which he had the honour to be one of the provisional Directors.

The chairman, after a time, summed up the matter very tersely. The Directors retired with somewhat drooping crests, and no harm came of this *fiasco*. But if those shrewd and industrious men whom the deputation sought to win to fill the share-list, reflected on the proceedings of the day, they might have readily come to the conclusion, that men who could convene a public meeting which they showed themselves so utterly unfit to control, or even to confront, were not exactly persons to whom might be wisely committed the expenditure of millions of money. The seed and the promise of years of floundering mismanagement were but too discernible at the moment when the plain-spoken Quaker consented to take the chair.

[34]Perhaps Henry Bosanquet, who became chairman of the company when it was established.

CHAPTER VII

Proprietary and engineering difficulties

Questions as to the occupation of land have often placed the servants of a railway company in circumstances of hostility to the county gentry, the occupiers of farms, or even the Court of Chancery itself. One instance may be cited which came before the latter tribunal as a question of contempt, but which led to the infliction of no penalty on the man who was unwittingly placed in danger of a summary chastisement. Negotiation had long been pending as to a certain estate, and either to bring matters to a crisis, or from some inscrutable reason not connected with any urgent constructive necessity, the Engineer-in-Chief gave positive written orders to the Sub-Engineer to enter the property, and to set out the line across it, in a definite and unmistakeable manner, by a certain date. The Sub reconnoitred. He found that the farmer was on the watch, with his own people and some supernumeraries, armed in various methods, in the full and avowed determination of resisting *vi et armis.* The young man took his measures accordingly. He determined to carry out his instructions, and at the same time to avoid a fight. He was soldier enough to know that this could only be done by a display of overwhelming force. He applied to the nearest contractor, who had a body of some seventy men engaged in the removal into embankment of the contents of a small turfy knoll, within three or four miles of the contested spot. The contractor, a stalwart and ponderous navvy of the old school, on whose rubicund and weather-beaten visage no amount of strong potations seemed to be capable of producing any change, entered into the plan *con amore.* At a fixed and early hour, the whole of the seventy workmen shouldered their pickaxes, shovels, and grafting tools, and, headed by their ganger, moved down the line of railway towards the scene of operations. The attendants of the Engineer, with theodolite, flag-staves, and chain, took another route to the same point. Their master cantered up on a pony in time for the junction of his forces. In the farm-yard, behind the gate and over the wall, appeared the barrels of some dozen fowling-pieces, the prongs of hay-forks, and other signs of the forces of the farmer, which, however, remained *perdu* at the sight of so formidable an array of invaders, and made no sign of the threatened opposition. The Engineer dismounted, and quietly ordered a hole to be cut in the hedge on the centre line of railway; and then proceeding along the indicated course, rapidly set out the half-widths with flags. The navigators, dividing into three bodies, cut a deep nick or "trig-line" in the centre, and turned off the turf for a yard and a half in width on either side of the land required for the railway. In an hour and a half all was quietly and thoroughly effected; the Engineer was on horseback again, and the workmen were on their way back to the cutting. Some proceedings ensued in London, of which the Sub was left entirely uninformed; and the only comment or compliment that he ever received for the service was, that on his next visit to the spot he found the sods cut out of the trig-line partially replaced, while some indignant owner or occupier had solaced his wounded feelings by cutting on the turf, in letters about a yard long and a foot wide, the inscription T FOOL. It was fortunate for the young man that the case was arranged, before the Lord Chancellor had time to add a grim Amen to his testimony.

The effect produced by the construction of a railway, especially when it passes the vicinity or the streets of a town in cutting, is by no means limited to the abstraction of a strip of land. The value of property adjoining the spot selected for a station is generally very largely increased. But residential damages are sometimes consider-able. The Birmingham and Gloucester railway ran through some fif-teen feet of cutting to the west of the town of Cheltenham. Trial shafts were sunk, which were stopped at some eight feet depth by quicksand. The meaning of this term, however, when used with reference to inlaid works, differs very much from the application of the same word to dangerous portions of the seashore. On the banks of the Solway, on the sands of the Pembrokeshire coast, and on many similar situations, are to be found banks of mingled mud and sand, often undistinguishable at the first glance from the hard and solid sands among which they are interposed, on which, if man or beast unwaringly step, they go down without remedy, and the treacherous surface re-closes over their remains. But when quicksands are met with inland, often as a source of immense expense and no small danger in tunnels, they are simply fine sharp sands, full of water. A natural basin or pit-hole has been formed in some impervious stratum, that has been filled by a fine silicious deposit, which, in the absence of natural drainage, is full of water. As you dip into the sand, the water replaces what you remove, and running from every side, brings fresh sand with it. There is but one safe and certain method of dealing with these subterranean quicksands, and that re- 49 quires patience. This plan is to tap them. The dyke which heads up the underground pond must be pierced. The best method is to lower a channel a few inches at a time, so as to drain the whole area with as little rush and disturbance as possible. By sinking two or three inches a day, the unstable sand is gradually dried, and becomes as sound a substratum for building as can be desired. It was in this manner that the Alstone cutting before described was attacked. A constant stream of pure water issued from the outlet, which was carried in advance of the face of the cutting; and the drain being gradually sunk to a depth of some two feet below the balance line of the railway, which ran up a gradient of 1 in 200, all the difficulty vanished in a few months. But with this tapping of a large pit-hole of oolitic sand, dammed up by a dyke of lias clay, came a shower of complaints. All the wells in the village had been dried. It was necessary to sink every one, from three to six or eight feet, as the case might be, steaning[35] the additional depth by stone laid in moss, and reaching again the reduced level of the water-line of the undrained quicksand.

[35]Lining with stone.

Long before this time a still more unexpected interference with ancient wells occurred in Kent. The canal which joins the Medway at Rochester to the Thames at Gravesend, passed through a lofty ridge of chalk by means of a tunnel of more than a mile in length; part of this hill consists of solid, compact chalk, but the lower portion of the duct nearest to Rochester was chalk with flints. When the canal was completed, the salt water of the Medway was admitted to fill it. But what was not anticipated by the Engineer was, that the wells for a considerable distance from the canal became brackish. The proprietors of the undertaking of course denied their liability. The case was tried at length, and it became evident that a sort of endosmose[36] had occurred along the lines of flint, the salt water of the Medway thus gradually but irresistibly making its way into the wells. The cost to the canal company was heavy.

[36]The passage of a fluid through a porous septum to mix with another fluid inside it.

North of the hill containing the quicksand, and alongside of the little stream of the Chelt, once a bright brook, but now the only sewer of the gay and populous watering-place, ran a belt of bog, a swampy, shaking morass. To drain this obstacle to the formation of the road exceeded the parliamentary powers of the Company; as it would have been necessary to lower the bed of the Chelt for a considerable distance beyond the limits of deviation. A plan was adopted which succeeded perfectly, although extremely simple. The whole surface of the field within the railway fences was levelled to the surface of the water, and covered with wattled hurdles. Turf was spread over them, and a couple of feet of sand was laid over the turf. The surface thus produced bore the railway without any

50

difficulty, and no inconvenience resulted from an obstacle which would have proved a source of great expense if otherwise dealt with. The wattled bog was only another instance of the application of the theory which gave birth to the macadamised roads of Great Britain.

The simple axiom that the soil will bear any weight, if it be protected from injury in detail, is what the French would call the mother idea of the Macadam system of road-making. To those who are acquainted with the noble *breccia* roads of Italy, it seems marvellous that the rate of travelling in England, the home and paradise of the horse as far as Europe is considered, should have been due to the dilettantism of a Scotch gentleman two generations back. Yet such is the simple truth. It deserves to be placed on record, in the very words that fell from the lips of that amiable and accomplished Engineer, Sir James Macadam, during a night's *tête-à-tête* in the Manchester mail.

Mr. Macadam, the father of Sir James, on retiring from commercial life, took up his abode in the vicinity of Bristol. He was soon placed on the list of country magistrates. During his former residence in Scotland he had indulged in a hobby as to road-making. The highways of the country at that time were in a state which can only be termed infamous. In some rare cases the mail route was paved, but as a rule the roads were in the condition of those country tracks or drives which have never been metalled, as it is now termed—that is, which consist of bare earth. With every shower the wheels cut deep into the unprotected subsoil. The roads became a series of miry ruts, crossing one another in every form of tangle. The guard of the Royal Mail, on seeing a waggon ahead, used to blow his horn to warn the waggoner to draw aside out of *his rut.*

In some particularly bad places large stones might be flung in to 51 give a bottom to the artificial bog, but here, as everywhere else, the weather and the traffic were too much for the surveyor; and it was regarded to be as impossible to make good travelling in the winter as to sail against wind and tide. In fact, the theory of a perfect road was one which, from the expense which it involved, would have confined serious road-making to metropolitan districts. It was, that the road must be built. It was to be a species of arch, resting on skewbacks or abutments, and thus relieving the underlying ground from the weight of the traffic. Thus, old English paved roads will be found particularly strong at the sides, the attempt being made to throw the thrust down into the earth by a series of wedges close by the lateral drains. In the old Roman paved road we find no indication of this unmechanical idea; the *Via Appia,* still to be found uninjured in some spots in the neighbourhood of Pozzuoli, being paved like the streets of Pompeii of the same date, and those of Naples at this day, with large quadrangular blocks of volcanic stone, laid as a simple pavement, but not at all as an arch.

It was against this theory, false in itself, and having the infallible result of rendering country road-making impossible, that Macadam waged war. "What do you want with an arch?" he said. "Keep the earth dry, and it will bear any weight that you can put upon four wheels!" And the simplest mode of keeping the earth dry, and preventing the formation of ruts, in which water invariably lodges, is to use small broken stone. A long series of experiments led the reformer constantly to diminish the size of the stone he recommended for metal. Flint or lime-stone, still better serpentine, granite, or hard trap-rock, broken into small cubes that would pass through a ring of three inches in diameter, was the material preferred by Mr. Macadam. Not that a cube of this dimension was in the state in which it formed the road. A further process of comminution had to be gone through, and, theoretically, the hammer of the road-maker should have continued work. But in practice it was more easy to leave this final work to be performed by the wheels of the carriages using the road; the crushing into fragments, and the binding of those fragments together by rolling weight over the surface, being thus conducted at the expense not of the constructors of, but of the travellers over, the road.

By degrees Mr. Macadam so far satisfied his fellow-magistrates and road trustees of the excellence of his plan, that the London road out of Bristol, as far as their authority extended, was repaired altogether under his direction. The success corresponded to the anticipation. Thus it chanced that one day the Postmaster-General, on looking over the winter programme for the timing of the mail, which

52   differed materially from that arranged for the summer, saw that there

was a stage for which no extra allowance was made. "How is this?" said his lordship; "you have allowed no further time for the stage into Bristol?" "No, my lord," was the reply; "the mail always goes as fast over that stage in winter as in summer." The Postmaster-General asked no further questions, none at least that elicited any explanation of so anomalous a fact; but he ordered his carriage, and made the best of his way to Bristol. Stopping at the post-house which bounded the trust administered by Mr. Macadam, he walked over the ground, and then, without saying anything further, drove back again to town. Then he wrote for the amateur commissioner to come up to him. "I want you to take the superintendence of all the mail roads of the kingdom," said the Postmaster-General[37] Mr. Macadam hesitated, and wrote to his sons, then making independent starts in life, to come to his assistance. The young men, who had generally their good-humoured jokes ready to pass on their father's hobby, now thought that the old gentleman was out of his senses. What! give up the actual business of life, and set to work at carrying out their father's road-making fancies all over England?! Reflection, however, and consultation, brought the younger men over to the views of the elder. The offer of the Postmaster-General was accepted, and when this account was given by Sir James, the income of that gentleman from the various trusts, which paid him £50 per annum each, was about £10,000 a year.

[37]This commission is not recorded in H. Robinson, *Britain's Post Office*, 1948.

CHAPTER VIII

The travelling destroyed by the railways

As the altogether unexpected results of the opening of the first sections of the Lancashire and Staffordshire railways became known to the public, the cities and towns that had driven the Engineers from their suburbs began to reconsider the matter. A line which now forms a part of the Midland system had been driven by local opposition into a course that was parallel with, although at some three to five miles distant from, the mail route. The coaches on this line were numerous and well appointed. Over certain stages, the most rapid travelling in the world was daily carried on. A distance of nine posting miles, or considerably more than eight measured miles, was traversed by several coaches within the half hour. The Hirondelle (called in the language of the locality the Iron Devil) and the Hibernia started at the same second from the rival coach offices. The frantic racing on the road which they had once carried on had been discontinued at the time of which we speak. No attempt was made by one to pass the other, unless a stoppage took place to pick up a passenger, in which place the second team rapidly took the lead. But in changing horses there was a stirring emulation. So close were the two parties to each other in their aptitude for this service, that as a rule the coach which changed at the last house got away before its opponent, so that the Hibernia would cross the Hirondelle at the changing station in the morning, and the reverse would occur on the return trip. The reason was supposed to be the infinitesimal loss of time caused by the attention given by the ostlers and stablemen to the proceedings of their rivals. In the case of the coach which stood behind, all their eager glances were directed forward, and did not sensibly interfere with the tightening of straps and buckles; but in the coach which stood first, the men were constantly glancing backwards or over their shoulders, and two or three seconds were thus lost; for it was an affair of seconds; twenty-eight was the average. Within half a minute of pulling up at the inn-door, each coach was again on its way.

All sorts of contrivances were resorted to in order to expedite the change. The Shrewsbury Wonder, one of the fastest and most regular coaches in the kingdom, or in the world, by the passage of which, men were in the habit of setting their watches, had the horses in some cases ready harnessed to the pole, and by the removal of a pin, and the loosening of the four trace-ends, the four incoming horses were removed altogether, and the four fresh ones slipped in with the new pole. In the arrangement finally adhered to by the Hirondelle, the new leaders awaited the coach with their faces towards it, and the wheelers were stationed on either side of the road, with their faces towards the leaders. The coachman unbuckled his reins as he came in sight of the inn. Pulling up with a sharp and steady check at the same spot daily, he threw the reins to either side, and walked straight down from the coach-box as steadily as if he were descending a staircase, and straight up to the new leaders. By the time that he had hold of their bridles, his own horses were unbuckled. Steadily holding their bits, he turned the two horses with their backs to the end of the pole, and then leaving them, ascended the coach-box with the same *aplomb* with which he had quitted it. By the time that he was seated all the attachments were completed. The reins were thrown up to him, a pair on either side, and in hardly more time than it takes to read the description, the Hirondelle was again in motion, Benton, the famous and steady whip, deliberately buckling the reins as he got up his speed to sixteen miles per hour.

The coachmen heard the tidings of the approach of their ruthless iron rivals with feelings that passed through the stages of incredulous contempt, of uneasy curiosity, and of alarmed enmity. When the Hertfordshire Vicar, before mentioned, threatened the proprietor of ''the coach'' which so often brought him from his Saturday night's whist parties to his Sunday morning's duties, with iron roads and steam coaches, Mr. Wyatt coolly replied, ''And do you know what I hears people a talking of? They means to send down here a patent cast-iron parson, to go by steam, and then they says they shall allas know where to find him.'' The retort seemed, in those days, as probable as the threat.

This confidence slowly deserted the Jehus. It may be supposed that the Engineers of a railway in construction near their route were not their favourite passengers. ''I suppose you'll set us up as mile-stones along your new steam road,'' said one. ''What do people want?'' said another; ''they will never be satisfied. First, they wanted ten miles an hour, then we gave 'em that; then they wanted to go faster, and we got up to sixteen miles an hour. 'I find horses,' says Mr. Dangerfield, 'you find whips, — that is your time to keep.' Now, they are not satisfied with that; they must go by steam at twenty miles an hour. By-and-by they won't be satisfied without 55 they go in less than no time at all. They will want to leave London at nine o'clock and get to Oxford at five minutes before nine.'' The honest coach-man little thought that he was a prophet. We do not yet travel at that imaginary rate, but our electric messages do.

When the range of the twenty-five to thirty miles that are within the limit of the comfortable daily work of a pair of well-kept horses is once passed, no mode of getting over the ground has yet been discovered to equal in the physical enjoyment it conveyed, that of the old first-class coaches. For riding, a distance of that amount is enough for exercise and for pleasure. If exceeded, relays are required; and riding strange horses is not generally agreeable. In making use of the now almost extinct post-chaise, the speed is comparatively low, unless four horses are put to. Even then the rattle and the swing of the chaise is not comparable to the steady roll of the coach, and the chaise has no coach-box. Then the change of vehicle at each stage is a nuisance; the turnpikes are a nuisance; the fees to post-boys and ostlers and helpers are nuisances, each of which you avoided on the coach, when the dues of coachman and of guard were according to an exact tariff. If you endeavoured to avoid the change of chaise by taking your own carriage, fresh anxiety replaced that which you avoided. The weight would be more than that of a post-chaise; or you would fear for your springs with reckless driving; or your movements would be fettered by waiting for your own horses to follow; or you would have to wait so long at some change-house while horses were not forthcoming, as altogether to defeat your object, if that depended upon speed. In fact, without the additional expense of a courier sent in advance, much time was lost by any method of posting, except on some of the most frequented roads.

But on the box of a well-appointed coach, on a fine day, over a good road, seated by a crack whip, or with the occasional charge of the ribbons, with four well-conditioned horses that knew their work, and their coachmen's mode of getting through it, travelling was a positive enjoyment. So fond were men of the road, that noblemen and gentlemen used to drive stages on the Oxford, the Brighton, and other lines. In one place a baronet regularly touched his hat for the coachman's shilling. In another, a gentleman who had not yet quite run through a fine property, used to drive his own four-in-hand some twenty miles to the scene of action, to drive the stage for a given distance, to return in charge of the up-coach, and finally to drive his own horses home. A major in the army, the near relative and presumptive heir of a noble duke, was a well-known and highly respected coachman on one of the Oxford coaches.

The stage between Cheltenham and Tewkesbury was one of the most rapid and agreeable run over by any of the fast coaches. As the hand of the town clock reached the quarter before six, the two opposition coaches, on each of which its coachman had been seated for four or five minutes, with elbows squared and whip advanced, while the guard stood behind with one foot on the step, would begin to move, rather with the gentle motion of machinery than with the ordinary jerk of starting horses. Steadily and rapidly the pace quickened to a flight, without a touch, with scarcely a sound to the horses. Then the bugle began to ring out a cheery tune, not the classical double note of the mail horn, but the merry strains of "Jim Crow". The bugle of the opposition coach, at some fifty yards distance, would re-echo. The fresh morning air, the fragrance of the wide hedgeless bean-fields, the distant rugged outline of the great Malvern range, clear in the early morn before you; the purple glory of the sunshine bursting over the Cotswolds behind; the steady, unswerving, rapid motion, combined to give a sense of exhilarating power for which the greater speed of the dusty, noisy, uninteresting train can afford no substitute.

The Shrewsbury, the Oxford, and the Brighton roads were among the most famous for their whips. There was a spot some four and a half miles from Cheltenham, on the Shrewsbury road, that served as a great test of the accomplishments of the coachman. Just where the red marl crops out from beneath the lias clay, a low hill rises, and runs parallel to the Severn. On the crest of this hill ran the direct road from Worcester to Gloucester, and the road from Cheltenham, running up this short hill at an incline of perhaps one in thirty, fell into the former route at right angles, at a spot named Coomb Hill. To drive up this ascent, and then along the crest of the hill, was easy enough. But to get gracefully down was the test. Most of the coachmen made use of the drag. Some would put on the drag and run into the hedge notwithstanding; but the steady and silent Benton, one of the best whips that ever handled the ribbons, seemed to take no further notice of the difficulty than was evinced by rather a firmer seat on the box, and a little more squareness of the elbows. Without a check, without an apparent effort, he brought his team round the square turn, and down the centre of the road, without the slightest change of pace, the coach rolling as firm and as steadily as over a straight bit of level ground.

Another style of driving, splendid in its peculiarity, was that of the Quicksilver mail. Over the undulating road from Plymouth to Falmouth two mails ran daily each way. One performed the distance at the rate of ten miles an hour, including stoppages; the other made the rate eleven miles, with the same allowance. The undulations might more fairly be called serious hills—Dutchmen would call them mountains. Great part of the road was got over at a three-quarter 57 gallop. But the peculiarity of the Quicksilver mail was the method of managing the skid. On dipping from the crest of a hill, the guard would lean forward over the coach, and, by means of a second chain adapted for the purpose, drop the skid before the wheel. Here and there he would have to try once or twice, but the sharp jolt of the off hind wheel, followed by a steady swing of the coach, generally told that his hand and eye were exact. In this operation there was little or no danger. But to take off the skid without stopping the coach required more nerve. Patent skids had long been in use, which the coachmen who travelled without a guard could raise and lower by means of a handle, and even without moving from the box. But to take off a patent skid it is necessary to back the coach. The number of stoppages that would have been necessary for this purpose on the Falmouth line would have seriously interfered with the speed required by the Postmaster-General.

The loose skid, with a light chain for lowering as before mentioned, was therefore made use of. To take off this shoe, the guard descended from his seat, ran to the off side of the coach, grasped the handle of the coach-door with the right hand, and thus accompanying the vehicle, seized the drag-chain with the left. Taking advantage of the first irregularity in the road that gave a swing to the coach, he snatched the skid from under the wheel as the weight reeled to the near side, and swung the link of the chain on its hook. Until you became accustomed to the perfect *sang-froid* and ease with which this service was performed, by a man whose ordinary function seemed to be to sit on a triangular stool behind the mail, and sound two notes on a long tin horn, the sense of danger and excitement was almost painful.

In the old French diligence travelling there was a greater number of horses, a greater amount of arbitrary power assumed by the conductor, more discomfort, less speed, and an infinitely greater noise, than on our English roads. Over some of the long heavy hills between Havre and Paris, or between Calais and Amiens, eight, ten, or even more horses would be attached,—you could hardly call it harnessed, to the ponderous and top-heavy vehicle, which after all would toil slowly up the hills, and come down them amid shouting and swearing, cracking of whips like the constant explosion of crackers, swaying from side to side, and the smell of the charred break on the wheel, which took the place of the skid. Travelling by a French diligence was a clumsy imitation of the English mail.

Probably the finest public travelling in the world out of England was, or is, if it be not yet displaced by the never completed Italian railways, that on the highway from Naples to Foggia, Bari, and Brindisi. The post-masters on this road are exacting and authoritative.

You have no voice as to the number of horses to be supplied to you. According to the class and estimated weight of your carriage, they are furnished at each stage, in a ratio calculated for the distance to the next change, and for the inclination or running facility of the road. Up one lofty hill, where the subsoil is clay, and road metal seems to be scarce, a carriage containing three persons is drawn by a couple of oxen. Four horses are generally the allowance for a rather lumbering barouche, but sometimes five, sometimes six are substituted. Over the rolling plains of Apulia, where the gravel affords a surface to the road equal to that of the finest English highways, a very good speed is attained. But the one point in which the Italian post-boys far distance their English brothers is the start. Clad in ponderous jack-boots, and cracking whips with only a less offensive fuss than the French (who are not born to deal with horses, while the Italians are) the postilions start on foot. They encourage and

scold their horses into a gallop; and, when they have attained their full speed, vault into the saddle, and continue their course in triumph. The Apulian post-boy and the Quicksilver guard would mutually have astonished one another.

CHAPTER IX

The repentance of the opposition

The country towns, which were enlivened and supported by the brisk and cheerful traffic of the road, began to look aghast at the menace of its decay. Where railways ran, it became more and more evident that ''long'' coach traffic would disappear. Local opposition, which had been unavailing to prevent the construction of the iron road, had only succeeded in diverting the new traffic from the towns. The coaches would be killed by the railway, and the stream of travellers would no longer pass through the ancient centres of business. The towns began to think they had made a mess of it.

A slumbrous old town, famous for the noble Norman windows of its massive abbey, and nestling in the frequent fogs of the rich meadows amid which the Avon joins the Severn, had succeeded in keeping the railway, which had been intended to pass through it, at a couple of miles distance.[38] To feed their own traffic, or, as they said, to accommodate the town, the directors of the railway sought to obtain parliamentary power for the construction of a branch. The mayor and burgesses, now as anxious to attract, as they formerly were to repel, the accommodation of the locomotive line, thought, that by opposing the branch, they might induce the company to make a deviation, and, at the expense of a slight additional distance, carry their main line close to the originally projected site. A conference was at last arranged between the railway authorities and the corporation; and a local Director, a tall, gaunt, harsh-visaged Quaker,[39] the Engineer-in-Chief, and one of his Subs, met the mayor, and two or three of his burgesses, to discuss the case.

Mr. Mayor was a character. A man of middle height, light-haired, red-whiskered, slightly bald, slightly inclining to corpulence, with a ready and extremely benevolent smile, and a quick, searching glance. Mr. Mayor was a solicitor, and no one knew whose title deeds were in his strong box. He exerted much control over a bank, and every one who might need banking facilities held him in respect—or at least in some awe. As an electioneer he was unsurpassed. He has been known to poll the exact number of votes set down in his final

[38]The town is Tewkesbury, and the tale given by Conder is not correct. Doubtless some of the citizens wished to stop the railway from coming there, but the line that passed it by two miles to the east did so because it was easier in engineering terms and could also serve Cheltenham, a much larger place, whereas if it had gone through Tewkesbury Cheltenham would have been placed on a branch. See *The Victorian City*, ed. H. J. Dyos and M. Wolff, 1973, i, 292-293.

[39]Perhaps Samuel Bowly or Charles Sturge—Quakers who took an important part in the promotion of the railway.

estimate, as made up at four o'clock on the morning of the election,—a very unusual amount of accuracy. His eloquence was highly popular. His accurate knowledge of the ins and outs of every one gave him great local power. Careful to avoid committing himself or his principal, he was yet fully aware that a fine Severn salmon, caught about the time of an election, deserved a price that was not to be reckoned at per pound. When one of the "Orange" public-houses sent in a bill for £180, and Mr. Mayor took the trouble to ascertain that all the excisable commodity that had been on the premises during the election saturnalia had not amounted to the value of five pounds, he did not allow the bill to be discharged without a mild remonstrance to the publican as to the propriety of being less prosperous in future. The house occupied by his worship was one of those comfortable, modest, substantial dwellings which we seem to have lost the art of constructing. Facing one of the principal streets, it was enclosed behind by a trim and lovely garden, rich with mossy lawns, and surrounding the apse and chancel of the abbey, so that the solemn strains of the organ, which had been used in the chapel of the Lord Protector, and which still bears the mace of the Commonwealth in lieu of the Royal Arms, fell softly on the ear as you walked beneath the larches and laurels. The house, as was fitting, possessed a haunted room; and the only thing which seemed wanting, as far as external appearances went, to the inmates, was that there should be daily service in the abbey. But Mr. Mayor was a dissenter, so that he passed lightly over this deficiency.

The mayor opened the conference with his frankest electioneering manner. The town much regretted the opposition it had given to the first bill of the company. In common with the cities and large towns of the counties traversed by the line, their own corporation had been ignorant of the real benefits of railway communication. They regretted that the traffic of the railway would suffer in consequence, and they were anxious to do what they could to support the line. Therefore they proposed, that the pending application to Parliament should be for power to deviate the main line and bring it through the outskirts of the town. To this the corporate support should be formally given. Of course the Directors would gladly avail themselves of the boon. For what would a branch be? A cost to the company, and a source of no emolument. An engine must be in waiting to meet every train, and perhaps not more than a couple of passengers would come over the branch in the course of the day. The mayor, therefore, hoped that all present would take the same practical view of the matter.

The Quaker Director replied. He fully agreed with the mayor as to the extravagant character of branch. No one could have put the case better. But he thought the wants of the town could be better provided for. The most convenient omnibuses had been recently invented to meet exactly such cases. Instead of having to make their way to a station, which must be at some distance from the greater part of the town, the omnibus would pick up every traveller at his own door. In little more than half an hour it would take him to an actual station on the main line, to the course of the through traffic. He was therefore so glad to find that the townsfolk did not wish for a branch!

The argument of the Quaker was a little too strong for the corporate authorities. "What!" said one of them, "We come to see if you would agree to give us a deviation, and you now propose not even to give us a branch, but to put us off with a 'bus!" So the conference came to an end, like more important conferences, without either party having been gainers by it, and the upshot was that the branch was made.[40] Greater silence and stupor seem to have fallen upon the ancient town, now that the streets no longer echo the pleasant rattle of the coaches, and the main traffic rolls through the wide bean-fields at a distance of two or three miles from the Abbey.

The instance above cited is but one of the many illustrations of the injury that was caused to the traffic of the country by the hot and improvident haste with which projectors planned, and Parliament sanctioned, lines of railway that were often extremely ill-adapted to the wants of the district. It is true that the idea at first prevailed among the promoters of new railways, that if a line once obtained the sanction of the legislature, its absorption of the traffic of the district was sure to follow; so that if it did happen to take a somewhat circuitous course, the profit on the mileage of the traffic would be so much the more. A selfish view of this nature is always short-sighted. Much of the ruinous loss incurred by parliamentary contests would have been avoided, if the natural course of the traffic had been studied in the first instance, and the railways had been designed with a wise forethought and a true adaptation to the actual requirements of the country. Such lines, as in the case of the London and North-Western, have a command of their respective districts, which is founded on the power of supplying their requirements in the shortest, and, therefore, the cheapest method. We have had serious lessons as to the relative value of short, direct, and well-planned lines, as compared with that of more ambitious and grasping systems, which depended for their traffic on that which they could divert from their neighbours. The unnecessary cost of loop-lines was an occasional result of bad selection of line. Thus the city of Worcester, a most important feeder of the Birmingham and Gloucester railway, was left at a considerable distance from the original main line of that

[40]Conder might have added that what Tewkesbury got was virtually a bus, for the branch was worked by horses from its opening in 1840 until 1844.

route. It was necessary to construct a branch. Engineering considerations made this branch diverge at an angle, far from a right angle, from the trunk. A second branch formed the first into a portion of a loop, or deviation; and then the main line between the junctions became useless.[41] Thus a double expense, and a permanent addition to the mileage of every train, ensued from the neglect shown in the first instance to the just requirements of the city of Worcester. It is impossible for the most ingenious speculators to control the force of physical laws.

It is true that, in the first attempts which were made to inaugurate a new system, the persons who would be the most extensively benefited by its operation were often to be found among its most active opponents. The authorities of Oxford thought that sound and religious learning could no more be expected to adorn their famous university, if the means were afforded to the undergraduates of running up to London in a couple of hours. Bristol and Southampton long carried on a ruinous struggle for the traffic of the West, each opposing the project of its rival, not on its demerits, but with the greedy and short-sighted design of monopolising the right of carriage. How very little the most sanguine projectors of the day could foresee the very proximate future, is abundantly shown by these and similar facts!

But if the true line was opposed, and the Act for authorising it was threatened, by either an ignorant or an interested opposition, it was an ill-advised course to evade such an opposition by taking a bad line. To wait for another session of Parliament, or for the result of that experience which was of necessity so novel both to engineers and to commercial men, was the only wise course. It is difficult to name instances in which that course was adopted, unless in cases where the original termini of lines in the suburbs of great towns were, at a subsequent period, replaced by central stations, as at Birmingham.

That very large sums of money, which were lavished on defeating the opposition of landowners, or on purchasing not only their land but their good-will, might have been saved by the exercise of ordinary prudence, was shown by the experience, at a date considerably later than those to which we have hitherto referred, of the projectors of a Welsh line of railway.[42] It was thought possible to connect the broad gauge and the narrow gauge systems of railway by a line which should open a district very sparse in population, but rich in mineral wealth. A direct communication between the commercial centres of Lancashire and the unrivalled natural port of Milford promised a through traffic of a magnitude that would render the line remunerative, even in the absence of any large amount of local business. But it was not, on that account, thought proper to neglect

[41]Not in the long run; as the service between Birmingham and Bristol was quickened the fast trains ran by the original line, passing Worcester by at a distance, as they still do today.

[42]The Manchester and Milford, which was authorised in 1860 to build a line from Llanidloes to Pencader, so providing a connection between the towns named in its title. The promotion of the railway, as described here, sounds admirably sensible. But Conder was evidently unaware of the result. Less than £8,000 had been subscribed for shares in the company by the end of 1863, towards an authorised capital of £666,000. Only part of the line was built, and the company was bankrupt from 1880 to 1906.

the local proprietors. One or two of the leading county magnates were called on in the first instance, and the advantages promised by giving such means of access to their estates were clearly shown, and, almost in every instance, pretty fully appreciated by these gentlemen. With their support county meetings were called at each of the principal towns on the route, before any serious expense had been incurred in the surveys of the line. The Engineer and the Solicitor attended, and explained in detail the local advantages that were promised to each vicinity. In one, the only manure used for the fields was lime. Even this had to be carted for a long distance. The farmers were not slow to appreciate the advantage of a depot in the middle of their district, where they might have not only lime, at a far lower cost than that which they actually bore, but stable manure, bone dust, or guano, each of which was simply inaccessible by road carriage. In another town it was shown that the residents would receive five per cent. on a £20 share, or at least would effect a saving equal to that amount, for every ton of coal which they consumed in the year. A man who burnt five tons of coal in the year would save more than £5 on this consumption, if the railway were constructed; and would thus receive a dividend on the £100 which he was invited to subscribe to that end, independently of any profit accruing to the company. In a third case ironstone, limestone, and lead ores would become productive sources of wealth from the mere circumstance of being rendered accessible.

These plain statements of undeniable facts went straight home to the county. Squires and farmers, tradesmen and shop-keepers, all saw that their interests were at stake. The pride of the county magnates was not affronted by ''a lot of Engineer chaps'' marching over their premises without warning, or dragging unauthorised chains over their lawns and shrubberies. Human nature, in a word, was enlisted on the side of the enterprise, instead of being, as in most cases, thoughtlessly hustled into opposition. The result of each of the four meetings was identical, with the natural difference that the more inaccessible the spot, the better were the terms offered by the proprietors to the projectors of the railway. In each case the large majority of owners of land to be traversed by the railway pledged themselves not only to offer no opposition to the bill, but to sell their land to the company at a fixed number of years' purchase on the rent; *and to take payment in shares,* or rather to subscribe for shares to an amount equal to that of the purchase money.

CHAPTER X

A West Country railway

A brief sketch of the history of a well-known English line of railway may be given as an illustration of the manner in which enterprises, in themselves most legitimate, were wrecked by their pilots at the very mouth of the harbour of departure. The principal actors in the scene are long since dead; and while the few survivors of a large staff will at once recognise both the persons and the scene of the drama, it is not likely that the recognition should be so general as to cause injury or annoyance to any one.

During the mania of 1835, an Act of Parliament was obtained for a railway from an inland town of great importance and activity, to a port with which it had much traffic.[43] The attention of the Engineer to whom such a service was entrusted might have been naturally divided between two routes, one of which, taking the most direct line, which was also that of the existing road traffic, passed through three considerable towns,[44] one of them being a cathedral city, and followed the natural indications of the physical and the social state of the country. An alternative line, however, was practicable, taking some distant but respectable towns in the route, and opening an agricultural country; not the thoroughfare of any important traffic, but passing an inland watering-place of even more importance for passenger traffic than was the terminal port which lay beyond it.[45] An alternative of this nature gives to the Engineer the rare advantage of being able to make a preliminary bargain with the land-owners; by informing them that the decision of the route may mainly, or altogether, depend, on the manner in which they will pledge themselves to the support of the railway through their own district, by entering into provisional agreements for the sale of the requisite land at fair agricultural prices, and for a proportionate subscription of shares. This precaution was in this instance, as in most others, entirely neglected. On the other hand, the ill-will of all the local interests was most ingeniously excited. The line was laid down on neither of the natural routes, but over an intermediate space of open country, far from any centre of population, requiring branches or other means

[43]This is the Birmingham and Gloucester. Its Act was passed in 1836.

[44]Bromsgrove, Worcester, and Tewkesbury.

[45]Cheltenham and Gloucester.

66

of expensive and inconvenient communication with its natural feeders, and finally terminating at the watering-place, though bearing the name of the port as its outward terminus. An amended Act of Parliament was designed, and applied for subsequently, for an extension; but the district in question was disputed and seized on by a hostile company, working on a different gauge.[46] The commercial line, therefore, ended in a watering-place, to which it ran through a country so destitute even of ordinary roads, that wooden bridges were erected for the crossings over the railway, owing to the inaccessible nature of their sites for carting down stone or brick for masonry.

The want of judgment evinced in the selection of the route may have been one main cause why some of the shareholders, when the cold fit first began to supervene on the enthusiasm of 1835, sought to abandon the undertaking. The Engineer, a man without practical experience of public works, endeavoured to strengthen his position by the aid of another Engineer of somewhat successful practice, who was then, greatly to the wrath of his own former leaders, rising to the commanding position which he afterwards long maintained, and in which he accumulated a large fortune.[47] The staff of the line was therefore ordered to prepare a revised estimate. The part of this estimate which it was considered most important to arrange was the sum-total, and for this end certain details of prices and quantities were decided on as normal; on which one of the Subs, who had seen both estimates and works before, thought it his duty to make a representation to the second in command as to their total inadequacy. That, however, he was informed, was no business of his, or of any one but the Chief, whose orders they were to fulfil. A staff thus converted into a mere class of arithmeticians entrusted with the working out of selected sums, of course produced the preliminary work required for the predetermined estimate. When all was ready, the Chief took charge of the consultation with his brother Engineer on the basis thus prepared. A period of some suspense followed, some of the young men, who had been invited from situations where promotion was certain, to take charge of actual works, feeling by no means pleased at having their situations perilled on work carried on blindfold. At last the report of the Consulting Engineer appeared, which was of a nature as little creditable to the profession as was the original—or rather, the revised—estimate. He, too, began at the end. ''The round sum stated by So-and-so,'' he reported, ''is no doubt, under proper management, sufficient to complete the line designed. But all the details are untrustworthy. This cannot be done for the price, and that is deficient in design. I can undertake myself to revise the whole plan, and to put it in the hands of contractors who will execute the line for the sum named, but I cannot, in any

[46]The Bristol and Gloucester, authorised in 1839, was built at first on the 7 ft gauge.

[47]The engineer was W. S. Moorsom; see Editor's Introduction, pp v–vi. The consultant was probably Joseph Locke, with whom Moorsom was on friendly terms. Locke left £350,000 when he died in 1860.

67

other way, support the opinion of your Engineer.'' It is a matter greatly to the credit of the *savoir faire* of the latter, that he continued to extract the honey of this damaging ''backing of one's friends,'' leaving the venom behind; and actually carried the day, and had orders to commence the works through the bean-fields.

But he was not content with this triumph, or with making the best of the services of the very respectable staff formed into the service of the Company. This included all the elements then available—a cautious officer of Royal Engineers, a working miner and mechanical Engineer, several *bonâ-fide* engineering pupils just out of their articles, a very accurate and beautiful surveyor and draughtsman, and other elements of a military but very appropriate origin. But a hint from a subaltern, especially if the subaltern happened to know anything on the subject (which involved his knowing more than his commander), always caused the latter to lose his temper; the consequence of which was, that he usually started on his own way, and only sought for advice when it was too late. As the details of the estimate had been adjusted to the total, the next and more difficult task was to adjust the execution of the work to the details of the estimate. Even this, however, was to be effected, to a certain extent, by ingenuity. No man could be responsible for the future, and as to what might occur a few years later,—sufficient for the day was the evil thereof. It would, however, make things pleasant for all parties for the Engineer to be able to announce to the Directors that the works of the line had been set within his estimate.

How far this gentleman did or did not deceive himself in the matter is one of those points on which it is neither easy nor necessary to arrive at a conclusion. The line of argument adopted, and the line of conduct based upon it, was certainly eminently calculated to deceive any who might put their trust in its rectitude. ''Great works,'' argued the reformer, ''are executed by great contractors, who make great profits by their execution. The great contractors sub-let the work to little contractors, who execute them for smaller prices, and the difference between the letting and the sub-letting prices constitutes the profit of the first contractor. I propose to secure this profit to the company, by sub-letting the works in the first instance. I shall prepare small sub-contracts for each individual work, and further divide the labour among different groups of tradesmen. I will contract with the quarryman for stone, and with the carriers for bringing it to the works. I will then sub-let each culvert and bridge to a competing master mason, who may, or may not, as the case will turn out, find lime, sand, and carting, as well as labour. Another contractor will erect a post and rail-fence; another will excavate the ditches; a third fencing contractor will set and weed the quicks. Earthwork

I will let a cutting at a time; and to encourage the small earthwork man, the Company will advance him 90 per cent of the value of his plant, to be deducted from subsequent fortnightly payments. Over all this vast and economical system the Engineers of the Company will keep watch. What else are they paid for? Idleness is the worst of dangers for young men.''

Among the practical disadvantages of this novel scheme, (apart from any consideration of the impossibility that such a complex system of bargains, each, of course, subject to its own peculiar leak, could be brought into harmonious co-operation by any less active energy than the eye of the master, sharpened by the goad of self-interest), might or might not be ranked the fact, that the earliest portions of all heavy works are almost invariably the least expensive. This fact is self-evident to the workman. It might therefore happen that works should be *bonâ-fide* let over a large district at very low prices, without affording any criterion as to the ultimate cost of the whole bulk of labour to be executed. Thus, a *prima facie* justification of an estimate might be secured, at a cost which there might be better means of ascertaining by-and-by than in the first instance. Meantime, however, advance would have taken place over the whole line, and then—why then the completion of works at almost any price is generally preferred to their abandonment. So it fell out in the case referred to.

The first, or one of the first, of the new order of primary subcontractors, a little man engaged to make small beginnings of great works at low prices, might have been photographed in the character of the genius of the system. A very little, sandy-haired, dusty man,

with a customary air of feverish anxiety, which not unfrequently deepened into an expression of intense anguish or profound and hopeless despair. He was a man of many trades—a bit of a manufacturer, a tradesman in a small way, a tentative constructor of sewers, an occasional burner of bricks, a collector and salesman of geological specimens, and an eccentric and fervent preacher. The mode in which he sought for solace when in difficulty, by the aid of a peculiar method of scratching himself, in a manner altogether unusual in public, recurs to the memory when the name of the worthy is mentioned. Heart and soul he entered into the new sub-contract system, taking half-a-dozen of these engagements at a time; and when asked how he arranged his time so as to get through his multifarious engagements, he replied, ''Why, the first thing as I does every morning is, I takes and eats no breakfastes.'' So he promptly verified the Engineer's estimates by taking numerous contracts considerably below their very moderate prices.

As time elapsed, however, this serenity became troubled. Sub-contract 70a was let at a rather higher price than the sub-contract 70, but for sub-contract 70c a much more considerable advance was demanded by the operator. The earthwork men, too, who for the first month or so went swimmingly on the strength of the advances made on their plant, considered the instalments of repayment to the company to be altogether intolerable deductions from the amount of the certificates and found themselves in prison, or in the Gazette, if they did not take the precaution of disappearing with their convertible means, leaving the mortgaged plant in the hands of the Resident Engineer. In the course of a few months great part of the works in progress was actually, if not always nominally, carried on as day-work under the inspection of the officers of the Company. The experience which these gentlemen thus acquired was no doubt valuable and extensive. It is questionable, however, how far the shareholders would have been content, could a statement as candid as the above have been laid before them, to concur in this mode of educating the rising generation of Engineers.

The natural consequences soon ensued. The works were either scamped, or carried on at an altogether disproportionate expense, or both. In some instances it was necessary, as time became pressing, to apply to *real* contractors, and to give them their own prices. Thus a cutting originally let at ninepence or tenpence per cube yard, was finished at the price of three-and-sixpence per cube yard. In such of the masonry works as had been advertised for competition, the lowest tenders were usually sent in by men belonging to the then newly-organised Trades' Unions. Protests on the part of Sub-Engineers were vain. Large bridges were let to these men, and the allowance necessary to keep a foreman or inspector on behalf of the Company, on each such work, was refused from motives of economy. The consequences were, that the daily rounds of the Sub-Engineer were watched, and that as soon as his back was turned, the scamping baffled description. The Engineer-in-Chief on one occasion unexpectedly visited one of these Union bridges, about an hour after his Sub had paid his morning visit of inspection. He saw enough to cause him to order half the work to be pulled down, but he did not see enough to lead him to change his system.

The southern portion of the railway lay through a district of lias clay, and a lofty neighbouring range of downs was composed of oolite limestone. Men familiar with masonry know that in many geological formations, the durability and building qualities of the stone of a given stratum are not questions to be decided by a reference to a geological map, or even by a casual inspection of the face of the country. Even the view of a working quarry will be far from satisfying either the Engineer or the Architect who is fit for his post. He will require to examine samples of the stone, from such quarries as may strike him as appropriate on careful examination; and will not be contented till he is fully aware how the stone in question has behaved in actual use, how it has stood the effects of frost, and how far it has been tested by lapse of time.

In the restoration of St. David's Cathedral, the well-known architect Nash availed himself of some beautiful sandstone from the vicinity, but such was the difference in the durability of different contiguous beds in the same locality, that a part of Nash's restoration is in as dilapidated a state as any portion of the ancient building; while the Irish oak roof, of great antiquity, looks clear and fresh, as if new from the hands of the joiner. In the range of oolitic hills to the west of Oxford, stone of every quality is to be found, from building stone —which, if quarried at the proper time of year, and duly seasoned before use, may be equal to Bath or to Portland—to an oolitic mass, composed of separate granules about the size of peas, and with no more coherence than a heap of sand. It is, in fact, a sort of fine oolitic gravel. It may well be supposed that proper care in the selection of stone was altogether incompatible with the sub-contract system, and frost and thaw left their visible autographs on many of the unsightly bridges.

The line lying through a clay country in the part where the works were lightest, and where the heavy unmetalled roads rendered the carting of stone for masonry a heavy expense, the simplest forethought would have prescribed the making of bricks, *in situ*, for some of the heavier works. This precaution was in no instance taken. It is possible that the Engineer-in-Chief may have derived a dislike to any intermeddling with brick-making from a little incident which took place in the early days of his experience.

A local landowner, who kept his eyes well open to the advantages which his own large estate would derive from the erection of a station, and the construction of a road leading directly to that station from the most fashionable part of the adjacent town, arranged for the sale of the minimum quantity of land requisite, at not very exorbitant prices; and fixed by agreement, the position of the station, and that of a first-class road. Then he made, to the Engineer-in-Chief, representations as to the necessity of providing bricks for the station buildings. Bricks were going to increase in price. It is a singular fact, confirmed by long experience, that bricks are always going up when one wants to buy, and going down when one wants to sell. If purchase were put off till the drawings were got out, the season would be lost, the price would be raised, the Engineer would be at the mercy of the brick-makers. A friendly interest, therefore, in the welfare of the railway company induced the landowner to counsel the Engineer to order of him (the landowner), the requisite number of bricks for the station buildings, which should be at once put in hand. The Engineer admitted the force of the argument, and the friendly nature of the adviser. He was a man of promptitude, and also a man of great accuracy,—that kind of accuracy which will work out a result to the third decimal of an inch, but which omits to check the painful calculation by the rougher measurement of the tape. It was fortunate

that on this occasion the Engineer did not foresee the need of any large quantity of bricks for his stations, because a large quantity multiplied by ten becomes a very large quantity; and while running out the estimate to the odd brick, the calculator unfortunately mistook the position of the decimal point. A fixed quantity, however, was bargained for, and the landowner commenced brick making.

Two or three months after the contractor had another interview with the Chief. "About those bricks," he said. "Well, the order is now in hand; more than half are made; a good many are in the kiln; I am ready to commence deliveries." "But I do not want any yet." "That is of course your affair; they were commenced for your convenience. The contract is quite regular; if you are not ready to take them, let me have something on account." "Well, that seems reasonable. The matter must be put in due course. You had better call on Mr. Blank, the Sub-Engineer, to measure up the work; he will send his certificate in the usual way to the office of the Resident Assistant; he will sign it and forward it to me. I shall countersign it and send it to the Secretary. He—" "But, my dear Sir, all that is very nice; you can arrange all these matters with your usual military precision. I do not understand your system, and have no wish to interfere with it. You will meet *my* wishes in the same manner. I have drawn on you for the value of a million and a half of bricks at three months. Just be kind enough to write 'accepted', and my bankers will spare me any further trouble in the matter, just as your Subs will save you."

What could be fairer? The brick-maker was more *au courant* at Bills of Exchange than the Engineer, who duly accepted the draft, and left the matter to take its regular course. The brick-maker took no further trouble. The Sub-Engineer heard nothing of the matter, and sent in no certificate. The Secretary made no payment, and the Bill of Exchange, at due maturity, came back protested, to the immense disgust of all parties to the bargain. Of course it was paid in time, according to the proper machinery, but explanations are disagreeable when they have to be made over dishonoured promises, even when written on stamped paper.

CHAPTER XI

A newly-formed staff

"One thing more", said the Chief, at the close of his first address to a newly-formed staff, "and that I have some delicacy in saying; but I think you will understand me, and I hope I shall not appeal to any of you in vain. I wish you to remember your position as gentlemen, as the officers of a great Company; and that your moral conduct and influence should be such as befits our relation to one another." It was good advice—advice, indeed, far better than was the example set by the giver; but it may be hoped that it was not thrown away.

The staff thus formed contained elements of both an effective and a very agreeable service. One or two marked characters, as often occurs, seemed to give a stamp to the whole of the body; for a regular mess was organised, served by a highly respectable sergeant and his wife. Three or four of the Subs and Draughtsmen lived in the house, in which certain rooms were retained as offices, and many dined or took other meals in common. An airy, convenient situation, on what were then the outskirts of a gay and fashionable watering-place, had been selected, and from the windows of the drawing-office a glorious prospect of fertile and varied country was bounded by one of the most picturesque mountain outlines to be found in any English landscape.

One of the eldest and sedatest of the men, a man who had been educated as an architect, represented what in the present time would be called the muscular element. He was a great advocate of training; he induced one and another of his companions to attempt their mile's run before breakfast; he had the gloves always *perdu* under the drawing-board. A pleasant companion, with some of the impulsiveness of the Creole, he was by no means averse to the relaxations offered by the gay neighbourhood, and prevented his companions from falling into mere mechanical routine.

This might have ensued from the undisturbed predominance of a man of a very different stamp, who had been educated as a surveyor, and whose precision of voice and manners, and extreme prolixity of work, had established his reputation for infallible accuracy. A quiet and amiable companion, his great desire was, first, not to be hurried; and next, not to be questioned. Thus a vague idea was established of his stupendous industry; and for a long time any computation from his pen was looked upon as twice checked, and absolutely correct. By degrees, however, it came out that accuracy and rapidity of work were not in inverse ratio. Those who had often thrown aside their own results, because they differed from those of the model computer, learned by-and-by the greater truths of the discarded calculations. But it was years after this time that, on hearing a discussion as to the reliability of the method used by this steady man in taking parliamentary levels, in which he never booked a reading nearer than the tenth of a foot, the Chief suddenly exclaimed, ''Why, that accounts for several very odd things that I could never make out in his sections.'' So it came to be understood that slow does not always mean sure.

Another man was a great contrast to the last: a tall, handsome Irishman, of College education, and whose abilities as a calculator or as an incipient draughtsman, had been rather generally, than specially, cultivated. He would have been decidedly ornamental, and increasingly useful, but for a weakness which had previously plunged him from a good station in society into one of the most anomalous conditions in which a gentleman could be expected to be found. A fair opening for restoration of his prospects was now offered him; but, unfortunately, ultimately thrown away. If he had possessed the power of self-control, and the virtue of keeping good resolutions, he would not have succumbed in the battle of life. One of the bad habits which he resolved and re-resolved to give up was that of snuff-taking. A certain stupidity of aspect, irritation of the eyes and nose, and imperfection in the intellectual operations which not unfrequently attended his morning appearance in the office, were attributed by himself to this unlucky use of snuff; so he boldly laid aside his snuff-boxes. There is a phase in the abandonment of any tyrannical habit, beyond which the reforming votaries seem slowly, or never, to proceed. In the case of snuff-takers, this stage is indicated by the abandonment of the snuff-box. But, as it is hard to break too suddenly with a use that has become a second nature, the discarded convenience is replaced, in this stage of the case, by a screw of whity-brown paper, usually kept in the waistcoat pocket. Thus the penitent satisfied at once his conscience and his craving. He ceased to consider himself as a snuff-taker on abandoning the outward badge of the fraternity. He procured the excitement, which he could not bear to miss, by exceptional pennyworths, which did not, after all, help to the desired result. After a time, people began to think that an honest box was preferable to a constant succession of dirty and untrustworthy wisps of paper. Later, however, it became evident that the cause of the irregularity in our poor friend's cerebral functions was not so dry as he maintained to be the case. When, in the course of professional business, he was, for a time, quartered alone in a country district, he sank into an unconcealed devotion to the least excusable form of drinking—that of endless and unlimited beer. Under that cloud he sunk below the horizon.

A fellow-countryman, however, joined rather later, who would drink beer only in default of wine, and wine only in default of whiskey. He was the genuine, dashing rattling Irishman of the novel; and if he had not passed through all the romantic adventures and hair-breadth escapes of one of Lever's[48] heroes, at all events he said that he had, and that, for a time, answered almost as well. He was a very good colourist, in every sense of the term; a fine, free-hand draughtsman, well able to get through a large quantity of creditable work, if he would be chained to the oar; a little man, with well-cut, handsome features, curling hair, growing back from his forehead, and that power of throwing his whole energy into the pursuit of the moment, which often gives men great influence over their companions. He could talk well, like most of his compatriots, tell a good story, sing a good song, mix a good tumbler of punch, make himself agreeable to either sex, and to almost any companion. The story never flagged for want of a personal link. *Quorum pars magna fui* might have been taken for a current motto. The remotest authority would be such as,—"When me fawther was dining one day with the Dook in th' Peninsuler Warr," or, "It was the way of all the gentry of County Clare in me grandfawther's time." Those who became acquainted with the adventures recorded in a certain kind of literature, came, after a time, to wonder at frequent coincidences, or to think it strange that the same individual should have been the hero of so many adventures.

One of the staff, the son of a man eminent in literature, and not unknown as a poet, took a dislike to the licence claimed by the newcomer. It suited the latter to remove this obstacle to his popularity. He softened his tone, he made some apologies; and taking a moment when the two were alone,—"Ah, Mr. So-and-so," said he, "many's the time I have heerd Mrs. Hemens[49] recating your fawther's poetry. Little I thought, last time I dined with her, to meet yourself." The well-aimed shot told—the singer triumphed; so far that the latter introduced him to one or two houses where he was intimate. The Irishman was asked to dinner at one of them, where he emulated at a modest distance the subsequently sketched character of "The Mulligan."[50] A day or two after, he caused the mess-sergeant to

[48]Charles Lever, Irish novelist (1806-72).

[49]Felicia Hemans, poet (1793-1835).

[50]I have not succeeded in tracing this character.

76

buff up a large pair of steel dragoon spurs; adorned with which, and with a light riding cane, he walked forth to pay his visit of compliment, and to complete the conquest he thought he had made on the preceding evening.

The last that was heard of this gifted but erratic man, one of those whose life proves that novelists do not either invent or excel the *dramatis personae* of actual life, was, that he had obtained some civil employment under Government—a survey was mentioned—in which he had opportunities of airing his spurs, to which he was said—it was, perhaps, a calumny—to have appended the further ornaments of a sword and a cocked hat. Let us hope that the "fond memory," of which he sung so melodiously, yet "brings the light of other days" around him.

It was at a later period that a portion of this pleasant group of associates was placed under the orders of a Resident, who had the reputation of being a practical man; a mechanical and mining Engineer, whose chief intercourse with his command consisted in making appointments, which he never kept, for the precise time of his next appearance. His main principle of business appeared to be the idea that any statement made to him was necessarily false, and that, as such, it furnished an indication of some fault which it was intended to cover. The poor man rattled vaguely from point to point over a large district of country, and sank at last beneath the duties of a post for which he was in no degree fitted. He once asked several of the staff to dinner. During the half-hour which it is sometimes difficult to pass on these occasions, one of his guests contemplated a well-known engraving that hung on the wall. "Yes," said the Resident; "yes, yes; very fine picture. It is Beelzebub's—Beelzebub's Supper; that's what it is—Beelzebub's Supper."[51] Dinner came, to the relief of all present, and a due gravity was happily maintained. A year or two later, two of the guests came upon a new delineation of the same scene in an Exhibition in London. "Ha," said one of them, "Beelzebub's Supper." A spruce, stout, oily gentleman stood by, in a suit of lustrous black and irreproachable white tie. "Belshazzar's," he interposed, in an authoritative and didactic tone of voice. His remark being welcomed with a hearty and irrepressible laugh, he seemed to collect himself, and to recall, with some hesitation, his Scripture history, fading from the dogmatic into a deprecatory and somewhat dubious tone, "Belshazzar's, *rather*."

On occasion of unusual pressure of parliamentary work, a local surveyor was called to aid the permanent staff. A man of mature age and quiet manner, he was at once one of the most industrious, and one of the best draughtsmen of his day. He had published one or two local maps, of great precision and beauty. He worked in

[51]John Martin's *Belshazzar's feast*, 1821.

spectacles—a very unusual phenomenon in an Engineer's office in 1836—and all his work resembled copper-plate engraving. He was curious in his instruments, his case containing not a few of his own invention; bisectors—a very useful and simple instrument; trisectors —proportional compasses, constructed on the same simple principle, but giving the instant division of a line into three, and thus often saving much time in construction; double pens, connected by an adjusting screw, for drawing, at the same operation, the two sides of a road or river, and thus at once saving time and improving the neatness of a survey; and dotting-wheels of various patterns. But the most memorable part of this surveyor's character was the manner in which, for three or four weeks at a stretch, he seemed able to dispense with rest, and almost with food. The regular office hours were from nine till five, with from half an hour to an hour for lunch. Working on there, day and night, this indefatigable man charged seven hours' steady toil as a day. Never eating a regular meal, only taking a sandwich and a cup of tea, or a glass of ale, while bending over the drawing-board, he thus contrived to cram three hours' sleep, in his clothes, and three days' work, into twenty-four hours. Out of the twenty-one days which a week can thus be made to contain, he made a charge, on one occasion, for seventeen or eighteen; and they were well-filled and well-earned days into the bargain.

The Chief under whom this industry was acquired was a man too full of contradictions to be dismissed in a line; perhaps difficult to be judged with fairness. A man of perfectly temperate habit, he had the physiognomy of intemperance. A very early riser, in his office at times soon after four in the morning, when, of course, he had no one to assist him, and thus worked at a disadvantage, he was so fatigued by the natural length of the day, as to be known to fall asleep over his own dinner-table at a party. His professional knowledge was limited, to use the most favourable term; but his chief hostility was evinced towards those who, in perfect good faith, would have enabled him to supply the defect. He desired to originate everything, but his inventions all proved lamentable failures. He was minute in calculation to an excess, but never took a broad view of his subject, or checked his prolonged decimals by a rough and truthful appeal to common sense. He soon involved himself in warfare with the Secretary of the Company—not an unusual mishap for an Engineer—but he managed to teach all his staff that they would meet fairer dealing from any one than from himself. He would occasionally, as became his position, blame any one of his staff for an error. But unless he did this on the spur of the moment—that is to say, with the loss of his temper—he was more likely to say nothing *to* the

defaulter, but much *of* the default where it might most turn to his disadvantage. He professed to be an eminently religious man; it is to be hoped that he was satisfied on this point in his own conscience. It is very likely that he was, for the nature of that faculty in him was so exclusively peculiar, that those who, professionally, knew him best, were unable to decide where it lay. With finer prospects than almost any Engineer of his time, with a lucky leap into a good business, and with a competent staff to carry out his duties, if he would only have allowed them to do so and gone to sleep, he contrived to drive away his clients, and to disgust his subalterns; to produce on every man who ever served under his orders the sense of injustice and of injury; and to inflict the most irreparable of these evils on himself.

The sketch of a service on which, with all its drawbacks, the three or four survivors may still look back with pleasant memories, should include, in order to be complete, some of the privates. The servants or chain-men of the Engineers were for the most part old soldiers, most of them sergeants or corporals; and from their habit of punctual obedience they gave great regularity and precision to the execution of all sorts of duty. Of these, one, almost the only private, was an Irishman of the name of Dempsey. Very deaf, very stupid, very ugly, and indomitably conceited, he preserved his place only by the sheer weight of his character—he had been so long known for a sober, honest, trustworthy man. According to his own account, he had seen much of the best service, and had been servant to Colonel Sir John Colburn,[52] to whose experience he always appealed as conclusive. He would fling open the bedroom door in the morning, perhaps half an hour before the proper time, and enter with hot water and brushed clothes, and a din and clatter that would have waked the seven sleepers. No injunctions would break him of the habit, and if you bolted the door, the battering of the injured and affronted orderly without, was such as to raise the whole house. He was afflicted with a mysterious and apparently incurable "intarnal disase," for which the faculty recommended rest and quiet, and that on no account should he touch anything stronger than milk and water. Consequently he never did—at least, so he said. He would at times appear to be stupefied, with flushed face, staggering gait, thickness of articulation, and increased hardness of perception. This was the effect of the "disase." But never would he be induced to take a glass of anything. On one occasion, a bitter day's survey in a frost, his master found a mug of warm ale, with a little gin in it, such a good antidote to the cold, that he pressed a glass upon Dempsey. The worthy consistently refused the tempting cordial.

But, after long groaning under a tyranny hard to endure, and harder to terminate, the Engineer to whose service the pattern Irishman

[52]Sir John Colborne, later Lord Seaton (1797-1863).

79

had been attached, found that worthy at last so undeniably and helplessly drunk, that the true nature of that "intarnal disase" became clear; and with that discovery the private had to retire into still more private life.

His successor, or predecessor, was a man of a very different stamp—a tall, handsome man, between fifty and sixty, brought up at a public school. He was the son of a gentleman who at the time in question was an Admiral in Her Majesty's Service, and a Knight of the Bath. Quarrelling, as a lad, with his parents, the boy ran off from home or from school and enlisted as a private in a cavalry regiment. He never could be induced to retract the false step; the family breach was never healed. Thus with the honourable feelings of a gentleman he grew up with the language and manners of a trooper. He did not rise in the English service above the rank of sergeant. As a trooper he often formed one of the escort of George the Fourth, in that monarch's journeys between Windsor and London, and used to recount how the four bays that drew the claret-coloured chariot, never breaking their trot, and never seeming to hurry their pace, kept all the horses of the escort at a three-quarter gallop, and kept the men up nearly all night in cleaning their bespattered accoutrements. The sergeant had, however, been in the service of the Queen of Spain, and had been lieutenant and adjutant in the Spanish Legion. In discharging his duties as a chain-man the sergeant was annoyed by rheumatism. Without saying anything to his master, he consulted a medical man, whom many people denominated a quack, and who was at that time becoming notorious by his attacks on what he called the "fallacies of the faculty;" that is, the regular practice of all the neighbouring physicians and apothecaries. One day, in walking over a large ploughed field, the Engineer found himself alone. He shouted in some displeasure, and after some time returned to look after the men bearing his instruments—level and staff, and chain. A field or two back he met the sergeant's son, who was also in his employ. "Where is your father?" "I cannot find him, Sir," said the lad. "I stepped aside into the next field, and I have been looking for him since." Anger gave way to alarm. After a short search the poor sergeant was found on his back in a furrow, insensible. He was conveyed to the nearest farmhouse, and on recovering there, confessed that, for some six weeks, the doctor had kept increasing his doses, while the pain which he called rheumatism rather increased than diminished with them. On the last visit, on the preceding day, the exposer of the "fallacies of the faculty" gave him a fresh medicament. "There," said he, "if you can stand that I shall be surprised: it is enough to kill a horse." It did produce that effect on a hale and sturdy man. He was conveyed home, but never again left his room, rarely his bed. Two Spanish books, his legacy to the Engineer on his death-bed, were the only memorials of a soldier of gentle birth and honourable feeling, who deserved a better fate.

CHAPTER XII

Some working details

In the construction of a railway, under the direction of an Engineer of such original and uncontrollable genius as was he, whose softened portrait will be immediately recognised by the few survivors of his once numerous staff, it was impossible but that incidents should occur of almost irresistible comedy. The one main feature of real business-like tact that kept the whole machinery at work, was the military punctuality which characterised the old soldier. This was precise and unfailing, and the acquisition of such a habit, as a second nature, was a counterpoise to much that was disadvantageous in the service. On one occasion the Chief directed one of his sub-assistants to meet him at the crossing of a certain road, with two men, at noon on a certain day, about a fortnight from the date of the note. A day or two after, the Engineer left the neighbourhood, having been suddenly summoned, it was said, by unexpected business to Ireland. The day fixed for the meeting arrived, without the appearance of the absentee, or any intimation of his return. "Of course he has forgotten his appointment," said the Sub to himself; "but as it is down in black and white, I shall just go to the spot, wait half an hour, and give myself a holiday for the rest of the day." He arrived accordingly to the minute. "I shall wait here for half an hour," he said to his men; "the Captain has made an appointment, but he is not likely to keep it." Likely it did not seem, but long before the half-hour had expired a dust was seen in the distance, a gig drove up at full gallop, and the Engineer, with his watch in his hand, sprang out with a brief apology for being ten minutes behind his time. He had not come in the gig all the way from Ireland, but he had sped in some incredible space of time across the country from Holyhead, perhaps inspired to some extent by the hope of dropping down on his Sub, for failure to keep an apparently forgotten appointment.

The personal punctuality of the Chief was seconded and fully equalled by the efforts of an assistant, to whom, as the real business of construction began to throw upon the Engineers the labour usually performed by the agents of the contractors, the former very wisely turned for aid. This latter was a man of system and a man of merit, and, more than either, actually an Engineer. The fag of his chief at school, his subaltern in military service, it was not now for the first time that he did the main part of the work, while the other obtained the lion's share of the credit and the pay. Without the exertions of this assistant, the whole work of the Engineer's office would have been involved in a whirlpool of irremediable confusion. Even as it was, the opening of the line was announced for a certain day, before a rod of roof or a yard of platform had been provided for at the terminus; and the Directors, at the eleventh hour, put the stations in the hands of an able and energetic architect, who very soon wrought a revolution in their condition. The exclusion of the Engineer-in-Chief from this natural portion of his functions seemed to be a just retribution, to those who were aware of an incident which had occurred some twelve-months before. An architect in practice at the terminal town,—the adviser, in fact, of our formerly-mentioned friend, the brick-making landowner—had been named architect to the railway. He naturally expected to have to prepare the *façade* and elevation of the terminal buildings, and to have a voice in the general arrangements, irrespective of those for the permanent way, which happened to be in cutting, at the spot. One day the Engineer addressed him, "Jerome" (let us call him) "you are the architect to the railway?" "Yes," said the other. "Well," said the first, "we want to put the stations in hand. The Directors asked me at the last Board if I did not consider myself competent to design them. Of course I could not deny that I did. 'Very well,' they said, 'we look to you as our Engineer, and we do not see what need we have of an architect.' You are our architect, and so, my dear Jerome, I fear that they will hardly call upon you for assistance." The adroit exclusion of a man who would have been content with a modest share in the architectural work, thus led, by a sort of poetical justice, to the introduction of a real and successful competitor.[53]

It would have been well if equal justice had rewarded the persistent and effectual exertions of the Resident before referred to. He had been stimulated by the hope of being appointed Resident in permanence. When the time came for the Board to decide on the appointment, the Engineer-in-Chief was equal to the occasion. "I have thought it right," he said, "to give you my advice as to the candidates before you. The first is, I have no doubt, fully equal to the duties of the position. If you had no one else you could rely upon him. You would not want a Consulting Engineer; but of course you would have to pay a proportionate salary. The second would take a lower amount. For all ordinary emergencies he would suffice. But he has no parlia-

[53]The architect of the station at Cheltenham is said to have been S. W. Daukes, a well-known local practitioner. Of the two architects mentioned here, was he the successful or the disappointed one?

mentary experience, no habit of dealing with men of business; he would depend very much on the advice of the Consulting Engineer. The third is, it is true, a very young man"—(he was just out of his short articles of pupilage to his supporter)—"but his salary would be very low; he is fully at home in my method of doing business, and as there are so very many questions connected with the line, which can be settled by no one but myself, whether I act as your consulting Engineer, or whether I have to be consulted separately as an independent professional man on each point, [I think] that you will probably find it the cheapest and best method to employ this gentleman." The Board took the advice, and their relations with their original Engineer enjoyed a protracted and vigorous longevity.

Originality of diction, charming as it may be when met with in society, has yet its disadvantages when it is incautiously made to displace the tedious and formal phraseology which is common to that borderland, disputed by the Engineer and the Solicitor—the specification of a contract. Among the numerous small contracts, or subcontracts, the superintendence of which left so little rest to the staff of the line we are describing, was one, or rather was a group, which placed the laying, ballasting, and maintenance of the permanent way in the hands of a man who, in a civil sense, was an extremely old soldier. Eminently respectable in his appearance, he managed to obtain credit for an unusual amount of ugliness as a token of moral worth and reliability. An impediment in his speech, which always became obvious at the critical part of the sentence, gave him time for reflection, and opportunity for swooping down upon his adversaries with unexpected, and often irresistible, force. He knew, by long and never-tiring experience, how to spell the word "scamp" in every possible manner, in Italian hand, in running hand, and in black letter. If he had been a few years younger he would have been a millionaire. The confidence which he would have inspired in a Board of Directors, or in a public meeting, if he had once attained so high a level as to be produced as an authority or a witness, would have been absolute. His fertility of resource was unfailing. Even when, giving way to the weakness of his nature, he became undeniably and intolerably drunk, he was never outrageous. He would gravely complain, at such times, of the terrible cold in the head, which had been brought on by staying on the works the whole of the previous night, or of over-fatigue caused by taking a patient to the hospital. He simplified the details of a very perplexed and ill-designed permanent way, by the original process of cutting the long holding-down bolts in two, sharpening the ends, and driving them into the sleepers with a sledge-hammer, instead of drilling holes and screwing on the nuts as per specification. As his extreme industry led him to carry on this bit

84

of blacksmith's work by night, it was a considerable time before his improvement came to the knowledge of the Engineers. With the mean jealousy which is so frequent a feature even of the scientific mind, they did not give the reformer credit for his simplification. They even drove him from his contract.

But long before this occurred, the Engineer-in-Chief paid a visit of inspection to the ballasting in charge of this worthy. The specification stated, with a *naïve* and original simplicity, that no bit of broken stone, to be used as ballast, should be larger than a man could put in his mouth. In the absence of either riddle, ring-gauge, or two-foot rule, the author of the specification flattered himself that he had thus secured a means of testing the faithfulness of the work. All was smooth and fair for the visit, all questions were duly answered, all faults seemed carefully avoided. But an extremely ugly man dogged every step of the Engineer. He crossed the line, the objectionable attendant did the same. He quickened his pace, he stood still, he spoke aside to the contractor, still the persecution continued. At last he could bear it no longer. "Amwick," he exclaimed, "what does that ugly fellow want?—is it you or me that he is dogging in this way?" "Oh, him?" said the layer of permanent way, with a grave designation of the obnoxious attendant with his thumb; "you mean him? You see, Captain Transom, that is my B—B—B—Ballast Gauge." He had drily selected the man of most capacious mouth to be found on the works, in case of any question as to the size of the broken stone.

If the chief set an example to his staff in precision, punctuality, and military promptitude, he might be thought to set an over-good example in another respect. There is no man who more deservedly earns, or more thoroughly needs, the rest of one day out of seven, than does the Civil Engineer. In his varied toil and responsibility, the demand made on both intellectual and physical energy, he finds within the limit of his profession the exercise of every faculty, and is thus deprived, to a great extent, of the solace of a hobby. If he works seven days out of the week, he soon finds it tell on him cruelly. The unusual harvest which the scythe of death has reaped in the profession within thirty years, the untimely death of Brunel, Stephenson, Locke, Rendel,[54] and so many others, can only be explained on the score of over-work. Night and day is all very well, if you get the Sunday; but fourteen days' work per week is certain, and not very tardy, suicide. At the same time, in all great public works, certain matters of necessity must be attended to on the Sunday. In tunnels, for instance, abandonment of work—at least as far as pumping, ventilation, watching of shoreing, and the like is concerned —would be in the highest degree dangerous and improper. In

[54]James Meadows Rendel (1799-1856).

85

hydraulic work the same rule often obtains. It is then quite consistent that a man should strenuously discountenance work on Sunday, and at the same time frankly attend to such particulars as involve the safety of workmen or of works. But our friend carried out his views on the subject in a manner peculiar to himself. He professed to touch no work on Sunday. He did his duty by attending to what was indispensable; but he satisfied that species of ingenuity, which took with him the place of conscience, by always dating such exceptional Sunday work on the preceding or on the following day. His Resident assistant, before described, once sent a mounted express to his Chief on a Sunday, to ask for orders in a case of sudden emergency. The messenger waited for the reply. In due time he returned with it at a gallop. The Resident opened the dispatch. His questions were replied to; the orders were precise, and they were dated—Saturday night!

The eyes of few men regard virtue without some degree of parallax. If it be true that one man's meat is another man's poison, it is no less true that about one man may linger a virtue which his neighbour may stigmatise as a vice. Thus what A may call a praiseworthy firmness of purpose, B may describe as a pernicious and intolerable obstinacy. Of this many instances could be given. One will be enough. The width of the bridges over the railway, to which so much reference has been made, was designed of the unusually narrow dimension of twenty-two feet. It was represented that this was an unwise economy; that the three-feet space between the inner rail and the centre of the railway was, in fact, a four or five-feet space on all occasions but that of a train moving with its doors open, and that for the sake of the permanent workmen on the line, the opening of the bridges should be wider. The Engineer was unconvinced. He built his twenty-two feet bridges. Not only so, but when one of them, a girder bridge, gave premonitory symptoms of falling, he inserted narrow cross columns in the centre line to prop the faulty girder. There was no room to splay the lines apart, as this was prevented by the position of the abutments. The parasitic columns were thus allowed to encroach on the centre space, while, from their narrow dimension, they failed to catch the attention of any one approaching at speed. So it unfortunately came to pass, soon after the reparation of the bridge had been effected, that a respectable and valued servant of the company, in passing under the bridge on a passenger train, while moving along the foot-board, came in contact with the column and was killed on the spot.

A year or so after, an Engineer, who knew all the facts of the case, was looking over some designs for a railway in the south of England, for which the same Engineer had made the designs. Among them was an exact reproduction of the homicidal bridge. ''I wonder,'' he said, with some warmth, ''that experience should give so little wisdom. Here is a duplicate of the very bridge which killed poor So-and-so. How could Transom have forgotten such a lesson?'' ''Do you know him no better by this time?'' replied an older man. ''Do you say he did it in spite of the accident? Not at all; he did it because of the accident.'' ''How do you mean?'' said the other; ''he is not a man to take a pleasure in killing the company's servants, even if he could do so with impunity.'' ''No,'' was the reply, ''but he wants to build this bridge, to prove that the other was right, and that the guard had no business to be killed.''

If any man ever chased Fortune from his roof, it was the Engineer in question. To many she comes once, to return no more if slighted; to him she paid visit after visit. It would have been so easy to secure her stay. She would have tarried with any man less obstinately bent on disgusting her; for he of whom we speak had many good qualities;—a frank military manner, open speech, punctuality, and love of order; domestic virtues, quick intelligence. For the scientific knowledge which he had not, there were those about him on whom he might rely. Nothing was needful, when the tide of success set in, but to keep quiet and modest, to increase his staff, to rely on his assistants, to pay them fairly, and to trust them with a faith that might have been reciprocal. It was from the lack of the latter quality, that shipwreck of a very fine professional position ensued. Borne well forward on the flow of the first railway mania, the same man was on the very crest of the wave in the second. He took in one year into Parliament more railway bills than almost any other Engineer. He lost them all but one, and that one was secured by the exertions of the experienced contractor who intended to make the line. All these lamentable failures might be traced to one source. It was a useless and blind perfidy. The principle of dating the Sunday letter on the Saturday, seemed to vex and distort the whole man. He would be unnecessarily and offensively mean to those on whose exertions, he learned too late, or perhaps never distinctly recognised, that his position actually depended. Very free in all his own professional charges, he was apt to economise in the remuneration of all those about him, nor could his promises, in respect of position and emolument, be ever relied upon. There are many men who will betray their neighbours for a consideration; this man seemed to take a pleasure in selling them gratis. Whether he did so merely to keep his hand in, whether it was a species of moral insanity, or whatever was the principle of his conduct, the natural result followed;—he became disliked, feared, and avoided. It is not wished to say anything that would give pain to any, if such there are, who loved this man.   87

The portrait can only be recognised by those who regard it from the business side. With them there can be no hesitation as to recognising it, no doubt as to the justice of the delineation. On the grave of a man whose reputation had been lost more hardly than it had been won, might have been written the epitaph, that its occupant lay there because he neither could nor would either learn or practise, the words ''shoulder to shoulder''.

CHAPTER XIII

A bit of work in Worcestershire

The casual visitor, or even the professional man, who observes the systematic manner in which the works of our great public enterprises are now carried on by experienced and competent contractors, can form little idea of what was the life of the Engineer, in those early days when all seemed to depend upon his exertions. The greater the detail into which the work was primarily divided, the greater the toil, and the more instructive the experience of the Sub. The men, too, who first rose from the ranks of labour to the position of master workmen or contractors, though far from dreaming in those days of princely estates in Norfolk, ducal palaces in Kensington, knighthoods, baronetcies, or Westminster Abbey,[55] were yet of a stuff more durable than some of their successors. Nothing was more remarkable than the instinct some of these men had for estimating. A man who could barely read, whose education in the matter of writing was merely such as to enable him to make a scrawl representing his name, or to prefix to this hieroglyphic the significant word "acep.," which did duty for "accepted," would walk quietly over a mile or two of country, and would then, by some mental process entirely unintelligible to the educated Engineer, name a sum for which he would execute the works comprised in the distance—a sum, moreover, that would come very close to the actual estimate of the man of science.

One of these native Engineers took a contract in this manner in Worcestershire. The site of the heaviest works was in a sparsely inhabited district. The provident contractor raised a little city of turf huts, and accommodated his workmen on the spot. Their daily wants were met by the ticket system, or what was better known as the "tommy-shop." Near the labourers' shanties was provided an emporium, furnished with all that their habits led them to procure. Ale in abundance, spirits, bread, meat, fat bacon, tobacco, shovels, jackets, gay crimson and purple waistcoats, boots, hats, and night-caps, were all to be obtained of the convenient "tommy-shop" keeper, who, though apparently an independent tradesman, was in fact a mere

[55]Perhaps referring to Somerleyton in Suffolk, where the contractor Peto began to build a great house in 1844, and to Albert House in London (now the French Embassy), purchased by the railway financier George Hudson in 1845. The first contractor ever knighted (in 1822) was Edward Banks, who had built Waterloo, Southwark, and London Bridges to the Rennies' designs. Peto was made a baronet in 1855. Robert Stephenson was buried in Westminster Abbey in 1859.

nominee and dependent of the contractor. Political economists would have been delighted at the ease with which the wants of the labourers were supplied without the need of money. Credit was always at their command. Even the newly-arrived and penniless ''navvy,'' provided that his boots were sound, could obtain a shovel on credit—that necessary implement of toil being all, except stout thews and muscles, that he was called upon to furnish. With a liberal credit, on the basis of still more liberal price, was combined a system of check, by way of tickets, which varied in different cases. Sometimes the whole onus of the account would be thrown upon the time-keeper. Sometimes tickets, say for five shillings each, would be issued to the workmen on account of wages, and the expenditure at the shop would be arranged, so as to exhaust these tickets by two or three settlements in a week. The pay was, generally, monthly. On a large sheet, the detail and the total of every man's time, calculated in quarter days, was entered by the contractor's time-keeper or chief accountant. The rate of wages, varying for earth workmen from 2s. 6d. to 4s. 9d. per day, was added. The total earnings of the month thus arrived at were entered in the next column. In the following one, came the amount drawn on tickets, or shop credit. The final column contained the cash balance due to the workman. It was extraordinary to see with how small an amount of actual specie the monthly pay was discharged. How much of that specie was to be melted into liquor within the next three days, was matter of very safe estimate. In fact, an argument by which the contractors met the desire of the Engineer, that the workmen should be paid weekly, or at the farthest fortnightly, was, that as they never resumed work till they had drunk out the balance of their earnings, it saved time and expedited the progress of the works to render these seasons of interruption as few as possible.

The profit of these tommy-shops was very large—often the main part of the gain of the contractor. It will be seen that, even in the country villages, small tradesmen manage to earn a living on the legitimate, or at least the customary, profits of their trade. If a sudden access of custom occurred, such a small tradesman would be in luck's way. If he was the general shopkeeper of the place, supplied every rural want, and found the demand for all his miscellaneous wares to increase in the same ratio, he would begin to make money. But if not only these advantages could be secured, but the actual amount of custom could be fixed beforehand, and competition altogether excluded, so that all check, either as to inferiority of quality or exorbitance of price should be remitted to the conscience of the shopkeeper, it is clear that something very considerable would stick to his fingers. If eight to twelve per cent. was the profit which a contractor, who

was his own foreman, could secure on the nominal cost of his work, the profit made by the shop out of the nominal cost price might amount to from thirty to forty per cent. In fact, in some instances, the poor men were shamefully ground down.[56]

Even unlimited beer at times palls on the taste. Men with a few shillings in their pockets are often disposed to seek amusements for the most part only readily to be found in great towns. While not given to dancing, like his Continental neighbours, the well-paid English labourer is not proof at all times against a weakness for female society, even if not of the most unexceptionable decorum. Thus, after a pay, some of the most hard-working, and, therefore, *pro tempore* richest, navvies might wander off to the nearest cities. They might not return, and their places would have to be supplied from the ever-circulating stream of inquirers for employment, that always fringes the course of large public works. Gangers (that is foremen) always like to keep their best hands—and will prefer receiving a repentant and penniless ''navvy,'' on his return from what they call being ''off on the randy,'' to taking on a stranger. With the ticket system another objection arises to the loss of a recently-paid workman—he spends his balance somewhere else. The shop, which has only made its profit on seventeen shillings and sixpence out of the pound which he has earned, being the amount represented by his tickets, loses the further profit on the remaining half-crown, and loses, moreover, the convenience of the return of that individual coin to the till. The contractor now referred to, bent his attention to remedy this evil. He did so in a manner that was at once simple, original, and effective. To prevent Mahomet from wandering in search of the loadstone in the mountain, he brought the loadstone to his

[56]These tommy-shops are sensibly dealt with in T. Coleman, *The railway navvies,* 1965, 79-82. The navvies were sometimes treated just as badly by local shopkeepers, however: see C. Whitfield, *History of Chipping Campden,* 1958, 213-215.

bivouac. He arranged for the presence of some of the objects of Jack's admiration—lodged them, clothed them, fed them, made them as comfortable and respectable as the circumstances of the case admitted, and set down a charge for their blandishments in the paysheet. It came under the head of "tommy-shop," and reduced, *pro tanto,* the monthly balance of coin coming to the pay-table. It will be clear to all those who think dealing with mankind is a mere matter of figures, that this contractor must have long since paid the debt of nature, for his name is not to be found among the members of the House of Commons. He never received the *accolade.* Nothing but a premature death could have prevented such a financial genius from arriving at the very summit of his calling.

The social economy of the contractor in question was, to a great extent, sanctioned by the example of the Resident Engineer. This was a man somewhat advanced in years; of good family and connexions, and not only endowed with a professional education, but a captain in one of the scientific corps of Her Majesty's service. He was a man of amiable disposition, a charming companion, precise in his military honour, but evincing the habits of an old campaigner, chiefly in his constant flow of original and terrible imprecations. Perhaps he had picked up the bad habit in Spain; but, certainly, every sixth word he uttered was an expletive;—ill-natured people said every third word. He seemed so unconscious of this peculiarity, that he was often known to reprove profane language in others. Whether he considered it a privilege of his rank is obscure. Even for the sake of illustration, one cannot venture to put on paper, in these more decorous days, a specimen of his conversation. It can only be delicately hinted at. One Sunday, his nearest colleague paid him a call. "Been to Church?" said the Captain; "I've such a — headache, that I said I'll be — if I go to the — church this — morning. So I took a little — of a walk, and what the — do you think I met? Why, three of your — men. I never heard — fellows swear in such a — way in my — life. I was — shocked. I said, 'Hallo, you —, is this the way you go on, coming from a — church?' I —'' But here the moral reproof which he administered veiled itself in such a perfect cloud of imprecation, that the modest dashes which have heretofore supplied the adjectives, altogether fail to indicate the fury of the storm. It was merely his way of saying that he thought it inconsistent for men to swear when coming from church. We must all agree with what he meant, however inappropriate was his unconscious method of easing his sense of propriety.

The peculiar originality of English expletive is a puzzling subject. The theological denunciations which, early in the present century, were thought no evidence of low breeding, at least, if ladies were

92

absent, seem so foreign to the practical gravity of the English character. It may be, that they were an inheritance left by the Puritans, and that the marked phraseology of the religious reformers who protested against the Cavalier licence of the Stuarts, has thus come down, in an inverted position, to our times. The idea and theory of evil language abroad differs much from our own. An angry Frenchman or Italian will have recourse to detailed impiety, to a purposed casting of offence at the invisible powers, that would make an English navvy shudder. In fact, those who utter such terrible words, with such constant tautology, seem to attach almost as little real meaning to them as do those who use them in the pulpit. It is a form, and a very unpleasant form, of speech; but the man who has just impre-cated evil on his neighbour for a chance trip on his corns, would be the first to plunge into the water to rescue him if in danger of drowning.

This harmlessness of intent in using very harmful language was curiously illustrated by our last-named acquaintance. He took up his quarters, at one time, at a farm-house, situated in an angle where a lane debouched on the high mail-road. A pet magpie was one of the establishment, a bird gifted with the abnormal intellectual activity and moral obliquity that characterise its elfin race. The magpie was hopping about the garden in front of the house, when, down comes a tramp along the road. "Mag, mag, mag," cried the tramp. "Let the bird alone, you unblest thief," cried the Captain, looking unexpectedly out of the open window. "I aint a unblest thief," replies the tramp. "Yes you are," reiterates the other, "an unblest thief; and you are worse, for you are a little, mean, sneaking, dirty thief, trying to steal the poor bird, with your unblest 'mag, mag, mag'." The man collapsed under the explanation.

There was a tunnel under the charge of this very particular Engineer. An irruption of water took place, and, as usually happens in such cases, the pumping arrangements were inadequate. Very frequently, the iron skips, which should have raised the solid contents excavated from the area of the tunnel, came up full of water. By some refinement of ingenuity, the contractor had managed to throw this particular risk on the Company, so that the Resident was in great request. Expense becoming beyond all bounds, in his fortnightly returns, he applied for orders, and was told to make a regular return of so many *cubic yards* of water removed from the works of the tunnel, at so much a yard. In all wet tunnels (and most tunnels are wet), the great desideratum is to tap the spring, and to run a drift-way or heading into the lowest portion of the tunnel, so as to let it drain itself. If this is unattainable, ample steam power, calculated for double the duty that the most liberal estimate of the supply of water demands, 93 is the first and least expense. The great cost of so many tunnels has generally arisen from inadequate provision for keeping the works dry.

In this case, the worthy Captain thought it necessary to bivouac, or to hut, for a few days on the spot. With the internal arrangements of his household his professional colleagues were not supposed to be necessarily acquainted. But the old soldier, though never shrinking from duty, did not like to lose his little comforts. So, after a day or two spent in solitude, his buxom young housekeeper was to be seen making his tea, morning and evening, and fastening comforter and cap for his evening excursion to the shafts. It raised a considerable amount of irreverent remark among the navvies. It was a union of self-abnegation and self-indulgence that told rather of foreign than of English habits. Brave old soldier, pupil of the Iron Duke! Those who could not avoid a smile at your peculiarities, never failed heartily to wring your hand. A noble old English name is on your tombstone.

CHAPTER XIV

A chapter on contractors

While the persons and the experiences of the Engineers who constructed the first English railways, and therefore the first railways in the world, may form the subject of many an interesting sketch, an attempt may perhaps be made, without need of much apology, to describe a few prominent characteristics of a very distinct body of men, by whose aid and energy the designs of the Engineers were for the most part wrought out. Novelists have indeed attempted one or two delineations of members of this class—but the imaginary contractor of the novelist is far from possessing the full flavour, the very pungency, of the representative contractors of actual life.

When it was first decided that the most feasible method of executing the heavy works necessary for the construction of locomotive lines, was by contract, a school or group of contractors had to be formed in almost the same impromptu manner as a School of Engineers. There were materials ready to hand. Special education, at least the education given by the schoolmaster, was less requisite in the case of those who thus represented the hand, than in that of those who might be regarded as the head, destined to create the new enterprises. The sturdy self-reliance of the English character, its practical, hand-to-mouth, mode of meeting difficulties, a keen eye to the main chance, and a readiness to carry a small amount of available experience to the best market, were the main qualifications that went to form a contractor for public works. Accordingly, the list of competitors was daily lengthened. London builders—keen, quick men, who had reduced their own branch of business to system, men, very often, brought up to the joiner's bench (by no means a contemptible school of handicraft), men who were familiar with the finance of the pay-table, who knew how far bankers would, or would not, consent to provide for the recurrent wants of the Saturday night—men who knew what could be done with bricks and mortar, who were well up in the matter of stone quarries, learned in lime and cement, ingenious in scaffolding—above all, able to put on the appearance of thorough mastery of the art of building, took the first Bank. Nor were these men metropolitan alone—Yorkshire masons, Birmingham bricklayers, miners, lime-burners, and quarry owners, often united the business of a builder with their own. Then there were the relics of the "navigators," who had dug our canals, who had gained experience in the construction of docks—the road surveyors and contractors called into existence by Macadam and by Telford. Sometimes a land surveyor would quit land-chain and jacobstaff,[57] to take charge of a body of workmen—later in the day a Civil Engineer would prefer the profit to be secured by a good contract, to the smaller certainty of salary or of fee.

[57] A surveyor's rod.

It was not until the Engineer became practically acquainted with the actual cost of the execution of works on a large scale, whether by the failure of the original contractors, which often threw the execution of every detail of work on the officers of the Companies, or by stepping in to undertake the entire responsibility of large contracts, when contractors did not happen to be forthcoming, that the estimates which he formed could become precise. Over and above the prices to be collected from the cost of small works, or from the price lists of various trades, a considerable amount had to be allowed for contingencies. Five per cent. was usually set down under this head. Thus, although the original estimates, according to which the Parliamentary capitals of the various lines were computed, proved, as a general rule, altogether inadequate, it was not because the prices were too low, nor was it because the designers wilfully shut their eyes to that which lay before them, but experience was defective. An Engineer who had to make a line between two given towns, might select his termini so as to avoid what he thought unnecessary interference with house property; the carefully-measured section from point to point would give a total quantity of earthwork and masonry, of fencing, ballasting, and laying permanent way, which could be duly estimated, and carried to account. Land, permanent way materials, a lump allowance for stations, and a percentage for engineering charges, would complete the estimate. Now, as far as these original estimates went, they were generally good. The contracts formed on the same data were generally let *below* the Engineer's estimates, but then came in the question of extras. The astute experience of the old "navvy" would often swell the list of these items, much to the amazement of the Engineer. "I will tell you a secret worth knowing," said one of the old Telford school of road-makers to a young Sub; "we've a maxim as puts a deal of money in our pockets —'The more you dissects it, the better it cuts up'." Now, the principle of lump contracts was intended to check this constant "dissection" of work, and consequent multiplication of extras to the benefit of the contractor; and the thirty years of contest between Macintosh and the Great Western Railway is the most striking instance of the difference between the statement of a final account drawn up by the Engineer, and that claimed by the contractor.[58]

Exclusive, however, of imaginary or exaggerated extras, which thus formed part of the price which the railway companies had to pay for the practical education of their Engineers, the items not contemplated in the Parliamentary estimates assumed a formidable magnitude. The heaviest measured item in an estimate (with the exception of tunnels or large viaducts), is the permanent way—a structure which costs £5,000 to £6,000 per mile. A railway of 100 miles in length might therefore be estimated to cost £600,000 for permanent way. But by the time that all the crossings, sidings, stations, goods stations, carriage sheds, engine sheds, and other accommodations for the traffic, each of which required to be laid with the same description of rails as the main road, were complete, from 125 to 130 miles of way, or even more, had to be provided. Then, as the traffic augmented in a ratio entirely unexpected, the inadequate provision made for the stations became evident. Furlongs of platform and acres of roof had not only to be provided, but designed. Some Engineers saved first outlay, and gained time and experience, by erecting temporary wooden stations, as was the case on the Great Western. Others attempted at once to provide for the wants of the future, and that in a manner which cannot be thought inadequate, although it looks poor and paltry by the side of our present lofty roofs. Thus the passenger station at Euston-square, with comparatively slight additions, has served for the enormous traffic of the North-western line. The station at Paddington—perhaps the most admirable station in the world, if beauty, convenience, and moderation be equally studied —is an instance of what was done by the Engineer who deferred the design of his terminus till he knew what was really requisite. The enormous waggon roofs of Charing Cross and Cannon-street probably err as much in unnecessary and costly magnitude, as the Euston roof does in over-economy.[59]

Besides way and stations, landowners and local interests swelled the amount of the bill. All kinds of unnecessary communications were demanded in sequel of the intersection of property by a railway, and each of these had to be provided for by bridge or other crossing, or commuted in cash. In the latter case the landowner invariably made the subsequent discovery that he could go round. Thus approaches to road-bridges, difficulties in crossing navigable waters, unsparing insistence by the Admiralty, the Ordnance, every Government department, every municipal body, every trustee, every turnpike trust, every one who had any *locus standi* for demanding outlay, on their pound of flesh, followed in endless succession. No wonder

[58]Two suits were brought by Great Western contractors against the Great Western company arising out of the work on the Bristol–Bath section of its main line in 1836–40: one by William Ranger, the other by David McIntosh. Both were extraordinarily protracted: Ranger's until 1855, McIntosh's until 1866. See MacDermot, *History of the Great Western Railway*, i, 54-55.

[59]There was much argument among railway managers, engineers, and the companies' customers about the merits of single-span arched roofs. The two named here, of 1864-66, were immediately followed by that at St Pancras (the biggest of all), then by those derived from it in Manchester and Glasgow. The plan was gradually abandoned; but still not without controversy, as in the case of the reconstructed Midland Railway station at Sheffield, completed in 1906.

that the railways of England have cost £42,000 per mile.

Of the five hundred millions which have been expended on the railways of the United Kingdom since 1830, it is certain that more than the half must have passed through the hands of contractors for construction. If, therefore, we allow that a profit of from twenty to thirty millions must have been cleared, by a body of men who have come into recognised existence within little more than the third of a century, we shall be within the mark. If to the actual benefits secured (however they may afterwards have been wasted) we add the power and influence natural to those who had the almost uncontrolled expenditure (so far as the pay-table and the bill-book goes) of three hundred millions sterling, it will not be matter of surprise that lordly estates should have been purchased, and noble palaces erected, by men who had known what it was to have to work very hard to make both ends meet.

The rich fund of comic incident which, if the tale of any great public work was fully told, would be found in the chapter that spoke of the contractors, is as yet almost virgin. But it is better worthy a place in literature, than are those Teniers-like portraits of the lower orders, that have succeeded to the popularity of Waverley, and that dispute the ground with such a writer as Bulwer, at once graphic and philosophic; for, in the gallery which the Engineer would collect, the pictures would not have the sole merit of similitude. In sudden elevation of unprepared men to stations of consideration, or of power the comic element is always mingled with a striking didactic moral. When tragedy supervenes, humour rarely is wanting. To hear a groom or a pot-boy talk, is an engagement only acceptable to the empty, or the innately congenial, mind. Set the beggar on horseback, and Shakespeare would not think it a waste of time to note his evolutions.

A grand feature of the mushroom growth of contractors, which has unfortunately disappeared in the unsatisfactory haze of "limited liability companies," was their loud and fearless self-assumption. A man who put a thousand pounds, in gold, silver, and copper, well counted, on his pay-table every Saturday night, was, in his own estimation at least, an important man. He would resolve to enjoy this achieved greatness without stint, and without loss of time. If other people failed to recognise it, he would show them their error. He would swear very loud, and fiercely regard his dependents; he would (behind their backs, be it understood) always call the Engineers by name, and express his contempt of their "theories." He would explain how *he* would manage the finance of the Company. Indeed, had it not happened that the money, often readily earned by railway contracts, had a most unusual propensity to make to itself wings and fly away, generally under the idea of rapidly doubling, trebling, or centupling its amount, the contractors' interest would, by this time, have held in its hand the fate of ministries, and filled Westminster with the din of an "Erie War."[60]

Look at one group; a large family, well known before the panic of 1848. The head was an elderly man, of large bulk, keen glance, and benevolent and confidential address. His education had been that of a joiner. He could read, even a paragraph from a newspaper, without much difficulty, could write his name, and, in case of extreme necessity, a line or two of hieroglyphics, to say, for instance, that he could accomplish some urgent and difficult undertaking "with grate E s." Finding himself in the present command of apparently unlimited resources, he saw no reason why his boys should waste their time in the acquisition of more "school-larning," than the very moderate amount which had sufficed him in his ascent to so proud a position. "Instead of bothering their brains with books, I sets 'em to look arter the men. Them's my hideers of headicashun." The eldest son, probably by that constant and unconscious tuition which clever men pick up by rubbing against their neighbours, had advanced far beyond the family level. A very tall, well-proportioned man, his light hair so receded from his temples as to show quite a Shakespearian forehead, which a pair of small blue eyes tended to magnify by contrast. A pleasant aspect, and an air which seemed invariably to inspire confidence, was only disturbed when he spoke; not that the extreme point and good sense of anything he said did not fully compensate for any originalities of grammatical construction, or discords of provincial tone; but the neglected state of his teeth betrayed, at once, want of culture and incompleteness of character. The next brother, a handsome, dark-eyed savage, with the stare of a Red Indian, or of a wild deer, would have been an able and successful man, had he been carefully educated. Beneath him, in various stages of hobble-de-hoyism, came two or three young giants, who seemed always outgrowing their simple garments. They only possessed the normal complement of limbs, yet, from the difficulty which they encountered in disposing of now an arm, now a leg, now a foot beating time against some one else's chair, now a large hand striking tattoo on the table, they recalled the structure of an Indian god, only they never sat straightly and compact, like those many-armed personages, but strode over chairs placed hindways before, or occupied two or three seats at a time, or in some other manner always contrived to fill a portion of space large enough to contain two or three meeker individuals.

The mainspring of the family fortune was the eldest son. He was essentially an able man. He was not unaware either of his own defective education, or of his own natural powers. His great weapon was a well-sustained secrecy; he never spoke till the right moment, and

[60]This was for the control of the Erie Railroad between Commodore Vanderbilt and Daniel Drew, assisted by John Fisk and Jay Gould; at its height when Conder was writing his book.

99

then he spoke to the point. He has been heard to say, that if he thought his hat knew what was passing in his head, he should burn the hat. He was a man who would take broad averages, and form large and usually safe inductions. How it was that he failed to establish a permanent position is even now a puzzle. He had large contracts at excellent prices; he was a good master, and knew how to select, and to attach to his service, good servants; he was modest in his personal expenditure, and steady in his habits. His activity as a contractor was in the golden age, when Companies paid the certificates of their Engineers in cash, and when Lloyd's bonds[61] were unknown; and all the semi-savage clan, up early and out in all weathers, formed a very good set of foremen. Then the way in which this man would command confidence was surprising. One cause might be, that he seemed never to ask for it, and rarely and sparingly to show it. But as his operations became more extensive, the credit placed at his disposal by the county banker, who became gradually implicated in his proceedings, was more than a respectable old firm could afford to lose, or even to have known. Both parties thought it better to widen the base of operations. Four or five banks within the sphere of action, would have been glad to share the amount, but the astute man of business went straight to the most impracticable banker of them all, the man of the most forbidding appearance, the most brutal to cross-question, the most keen, in his own opinion, to detect a "kite" by the smell. The aid of this very self-asserting manager was not sought, it subsequently appeared, until a time when no prudent banker, with the means (which a prudent banker should have possessed) of learning how matters really stood, would have advanced a hundred pounds to the contractor. Yet the man, who was the terror of the speculator, and thus applied to at the eleventh hour, figured as a large creditor in the balance-sheet.

A very different and very inferior man rose to temporary importance on the strength of a lucrative contract in Wales. What had been his special training was not evident; it could have been only of the most primary description, although he shared with the elder man, of whom we last spoke, a predilection for preaching on convenient occasions, or for taking the chair at some of those meetings for benevolent and religious ends, the conveners of which are apt to regard rather the subscriptions, than the orthography, of their president. The money which he pretty freely made he even more rapidly invested in a coal mine, from which the only combustible that he ever succeeded in raising was water. Now, it is true that water, if resolved into its elements, is the best of all combustibles; but that scientific fact does not prove the economy of spending thousand after thousand in nothing but extracting water from great depths in the bowels of

[61]Lloyd's Bonds represented a simple acknowledgement of a debt owing by a railway company to a contractor, under the company's seal. The contractor was enabled to use them as security for loans (at a high rate of interest) from a bank.

the earth. This contractor was more than usually ignorant, and, consequently, more than usually self-confident. "If So-and-so" (the Secretary of the Company) "can count on thirty thousand towards that 'ere 'undred thousand, he must be a pretty fool if he can't see his way clear to pilot the company through their difficulty," he said on one occasion. "He'd better ask me to do it for him. I shouldn't have no happrehension under them circumstances myself."

The helpmeet of this self-reliant miner was a woman who showed the superior tact of the better sex, in circumstances of sudden elevation, as compared to that evinced by men. Very portly, very fresh-coloured, rather pretty, she never aspired beyond a smart bonnet and a pony-chaise, and thus set a notable example to many not better fitted than herself to indulge in more brilliant luxuries. She would take her turn in seeing that the pot was kept boiling, look round the stable-yard at odd, but useful, moments, or descend upon her lord and master, if his visit to a public-house was too prolonged, and bear him off to conclude his discourse in domestic privacy. Matters would have gone better if the grey mare had been always the leader of the team, or rather of the tandem.

A speciality of many of these people was their intense and inexplicable angularity. Their elbows seemed always *de trop*. An Engineer had been sent into Cornwall to report on some granite quarries. On the line of the "Quicksilver" mail he saw a post-chaise advancing to meet him. From the window of the chaise protruded something, which on nearer approach proved to be a man's elbow. Why that feature should have at once seemed to announce its owner, may seem inexplicable, but it did recall to mind the worthy miner, who was thought to have been hundreds of miles distant. But as the vehicles crossed, the elbow disappeared, or rather was replaced by an arm, terminated by a large and unwashed hand, which gave a patronising wave of recognition, accompanied by a nod from the unmistakable physiognomy of the enterprising contractor.

The reader may, not unnaturally, look for anecdotes regarding a member of this active fraternity who, subsequently, during his passage to a disastrous collapse, assumed the prominent renown of a national benefactor, a religious founder and philanthropist, and a not untitled legislator.[62] Many instances might be given of keen practice and prompt resource on the part of this well-known personage, but they are, as yet, too recent to be unobjectionable in print, at all events, in pages which, while containing none but hitherto unpublished personal recollections, have endeavoured so to refer to all matters of a deprecatory character as to avoid creating any painful feeling in survivors. Those whose memory is hardly, although justly, dealt with, have been spoken of in their professional relations alone. The

[62]This is Samuel Morton Peto, who was a prominent Baptist, MP 1847-68, was made a baronet in 1855, and crashed in 1866. Conder's phrasing here implies that he took an unfavourable view of him. There is an important book to be written about him, assessing him fairly.

few survivors who have borne a part in those relations, will at once recognise the truth of the tableau. But no relative or representative, who was familiar merely with the private life of the subjects of these reminiscences, will be likely to know to whom they refer.

A Civil Engineer had been invited to take a large contract, in a part of the country very remote from the centres of population, at highly remunerative prices. On his way to inspect the locality, he happened to meet, in the coffee-room of one of the railway hotels, with a contractor of large experience and heavy engagements. "I shall be glad of your advice," said the Engineer; "I have such an offer, which seems to me to deserve much consideration. There is certainly ten thousand pounds to be made by the acceptance of it, possibly more. And the reliance that is placed on my character by the Engineer, who says he wishes for a man whom he can fully trust at such a distance, weighs for very much in the scale." The contractor heard all, put one or two questions as to leading prices, and then repaid the young man's confidence. "I should never recommend any one to take a contract," said he. "But example is better than precept; and I do not mind mentioning to you in confidence, that my object in being here is, to make arrangements to get out of all my contracts. I shall have enough to live upon, in a quiet way. I am not fond of speculation; and I shall prefer a modest certainty to the prospect of very large profits, with all the trouble and anxiety that are involved."

If this advice was candid, the giver must have changed his mind with unexampled promptitude; for it soon after became known that, so far from retiring himself, he had just bought out his own partner, and that, as a preliminary to taking a larger contract than had been, up to that time, confided to the exertions of any single contractor.

CHAPTER XV

Engineering "Works"

The traveller who, for the first time, enters what is emphatically and appropriately called the Black Country, may be excused for the fancy that he has, by some mistake, actually entered on the gloomy regions which have been described by the verse of Dante.[63] In the days of the old coach traffic, the slow approach to the district of smoke and of flame gave warning to the passenger of his arrival in these gloomy regions, and produced an impression of wonder and of dismay. But by the rapid course of the locomotive, a sudden change of scene is now effected which is even more striking to the imagination. A traveller engaged in conversation, or in reading, lifts his eyes, as a glare falls suddenly on his page, and beholds a pair of lofty cupolas vomiting flame. All around, the earth is black; the hideous, unarchitectural buildings are black; the half-stripped men, toiling to fill the ever-devouring furnaces, are black; the sky is a dense canopy of smoke, glowing and angry with the reflected light of the undying furnaces. The blast, urged by the steam fan, emits a constant and penetrating hum. The cranks and wheels of the gaunt, skeleton-like, steam-engines, working without shelter and without rest, raise a dismal clatter. It requires a strong admiration for the principles of political economy, to enable an observer to enter the manufacturing district without an extreme feeling of repulsion.

Under these gloomy skies, and amid this carbonised landscape, is to be found much of the backbone of the English character. Habits of skilful enterprise, of energetic toil, of helpful self-reliance, character-ise the inhabitants of the manufacturing districts. But the good is not unbalanced by the evil. Thriftless waste, profligate destruction of the precious fuel, of which, after all, the quantity has a definite limit; rough, rule of thumb, mode of work; brutal ignorance; savage disregard of the decencies of life; drunkenness, elevated into a species of hebdomadal worship; can hardly be said to divide class from class in these districts; for the masters, in most instances, are but richer or cleverer workmen.

This is not always the case. In the Yorkshire and Lancashire valleys

[63]At this point Conder left railway employment to follow his former chief Charles Fox into contracting. This chapter and the next reflect his experience of Fox's works at Smethwick, outside Birmingham.

there is much more education, much more refinement, than in the Staffordshire pandemonium. You may accompany a rough man, in a home-spun coat, to his home; music, flowers, books, a neat conservatory, a choice of piano, French journals on the table, prepare you to be introduced to well-educated, or even to well-bred women, and to a peculiar mode of life that unites, in a high degree, utility and civilization.

In the great Welsh coal-field, where the centres of population attest the natural abhorrence of the steam-engine for the picturesque in scenery, the hand of man has not so far obliterated the features of nature as in the central coal-basin of the Midland Counties. The vapours poured forth from the smelting furnaces of Swansea, poison the vegetation for many miles around; but the Welsh hills are imposing, even in their nakedness.

As you ascend the busy valley leading inland from Cardiff, wood and pasture, noble seats and picturesque steeples, are interspersed with the discoloured patches that mark the immediate vicinity of mines or iron works. But the beauty of the Belgian coal-fields, the neatness of their industrial villages, the economical consumption of fuel, leaving the sky as pure as in the agricultural districts, will be vainly sought in any seat of English mining or metallurgic industry.

If you enter one of the great works, you become impressed with a sense of the energy of self-taught labour. You find before you a large quadrangle of low brick buildings, with roofs of slate on iron principals, or covered with corrugated iron. In the centre of this quadrangle is the boiler-house, the lofty chimney giving sign at once of the activity of the establishment, and of the neglect of a wise economy of fuel. Turning to the right after you enter the great front gates, you see at the end of a corridor, into which small rooms used as offices open on either side, an apartment with glass folding-doors, and containing, almost filled by, an enormous desk. Standing behind this ponderous piece of furniture, or perched upon a horse-hair stool, is a short, stout man, with a fresh colour, a quick black eye, a bullet head, and dressed in a suit of black, with swallow-tailed coat, low shoes, and white neck-tie. If your business be with this active and energetic controller of the works, you soon learn to understand the genius of the place. You may find him a self-educated Scotchman, shrewd, keen, not unkindly, but of an essentially combative disposition. He will oppose everything you say, from sheer force of habit, oppose you if you say nothing, and break off in the middle of the conversation to roar out a paragraph of a letter, to be written by one of two or three clerks, who, perched on stools or boxes, wait always in anxious readiness to note down the efforts of the inspiration of the moment. When you become accustomed

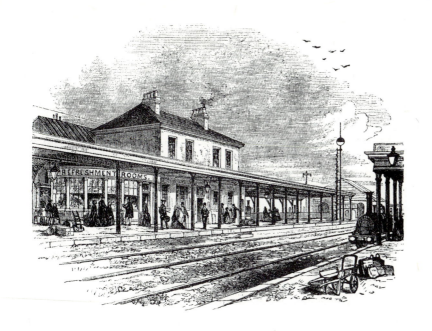

to all this sort of thing, you understand how the native energy of
the man, developing itself in every movement, makes itself felt
throughout the entire establishment, and while often producing confusion, often causing loss of time, waste of material, blunders of all
sorts, yet keeps up a vigorous life and motion. Unfortunately, the
worthy manager, during the intervals of his occupation in early life
at the carpenter's bench, heard some account of the number of secre-
taries for whom Julius Cæsar found occupation. He thus came to
the conclusion that the grand test of the capacity of a man of business
was, to be able to dictate four letters at a time. It is true that, in
the daily attempt to rival Cæsar in this respect, a paragraph intended
for A would sometimes find its place in the letter dispatched to B.
The lengthy and verbose productions of the four clerks were generally
about four times as long, twice as confused, and three times as unsatisfactory as would be the correspondence of a single able writer. But
the correspondence went on, after its fashion, amid orders to small
tradesmen, fierce bullying of foremen, civilities to visitors, attention
to many real demands on charitable kindness, an unsystematic, never
satisfying rush of occupation, only kept together by the one bond
of the personal energy of the manufacturer. He did his work loosely,
unsystematically, wastefully; but with all that, he *did it*. Railway
wheels and fittings, trucks and horse-boxes, lofty slip roofs for government dockyards, working stock for foreign railways, bridges, and
fittings for fire-proof buildings, columns and girders, rafters and ties 105

for a vast structure of iron and of glass, all found their place in the never-ending correspondence; all found their way, in time or out of time, from the wharf at the corner of the works, by canal or by railway, to their various destinations.

You may like to take a glance round the works. The boiler-house, as before mentioned, stands alone in the centre of the quadrangle, but the engine which it supplies with steam, and which provides the whole driving power of the establishment (apart from the driving power of the manager), will be found in the central building, opposite to the main entrance gate. Here it sets in motion polished steel shafts, which run the whole length of this side of the quadrangle; from drums, keyed upon which, bands are carried to the numerous planing, drilling, punching, shearing, and screw-cutting machines, which, each attended by a careful fitter, and one or two labourers, fill a large central hall in two storeys. To the extreme right is a lofty building with towered roof, under which are to be seen a couple of cupolas, and outside of which is the open grate of an air furnace. The loud, steady hum of the steam fan is most audible in this part of the works, which the floor of black, fine sand at once points out to you as the foundry. Here you come in contact with a set of the wildest, the most grimy, the most independent, and, unfortunately, the most drunken and troublesome of any English workmen who have any claim to the title of "skilled." They are the moulders. Some excuse must be made for the exhausting nature of their occupation; but, after all, the labour of the moulders is far less severe than that of the smiths, while the latter form by far the most respectable, as well as the highest paid, of the teeming population of the works.

If you enter the foundry in time to see some large casting run from the cupola, you will witness a scene which seems ever attractive, even to those who are familiar with the fierce energy of molten iron. The outline illustrations of "Fridolin," or of the "Song of the Bell," although drawn from German workmen,[64] are yet the best representations of the labour of an English foundry, with which we are familiar. The foreman of the foundry removes, with a long pole, the plug of clay which closes the orifice of the cupola, and a steady, limpid stream of dazzling white metal pours out, shedding a clear light through the gloomy shop, and every now and then exploding in little fountains of sparks, blue, white, and red. The stream is received in an iron pot, lined with clay, and either supported by two long handles, composed of wood and iron, or slung by a chain from the arm of a lofty crane; two or three of which so communicate with each other as to enable the heavy vessel of molten metal to be slung to any part of the foundry. When brought, by crane or by hand, to the spot where a sort of funnel has been carefully formed in the

[64]The reference must be to an illustrated edition of *Das Lied von der Glocke* by Schiller, 1799, and—perhaps—of Gottfried Keller's story "Die Leute von Seldwyla" in *Die drei gerechten Kammmacher,* 1858, in which there is a character named Fridolin.

sanded floor, the iron ladle is slowly inclined, till the contents, which assume a redder hue as they begin to cool, flow into the funnel, and gradually disappear under the floor. That the iron is not lost, and that some plastic operation is going on beneath your feet, is soon apparent. A slight explosion, a mere pop, is heard, and the floor of the foundry, for some distance, is suddenly covered by an illumination of fairy blue flames. The heat of the metal has resolved into its elements the water mingled with the sand. The oxygen has attacked the iron, the liberated hydrogen, making its way through the sand, burns, so soon as it comes in contact with an inflammatory substance, with a steady flame of low illuminating power, but intense heat.

Corresponding to the foundry, at the left angle of the inner row of buildings, is the smithery. The labours of this department are now much reduced, as far as human exhaustion is concerned, by the introduction of the steam-hammer, that wonderful application of mechanical skill, with which the inventor, Mr Nasmyth, could crack a nut without injuring the kernel, or could forge the heaviest anchor. Before the introduction of this admirable tool, one of the finest sights in an engineer's factory, was the forging of a large piece of iron. In a large smithery there might be as many as sixty-four forges, grouped together in nests of four, each around a central chimney. In the Government smithery at Pembroke dockyard, no chimneys were apparent. Low, wide, open forges were set against the wall of the building, the tuyeres, hollow, and kept cool by a circulating flow of water, were urged by a blast from the steam-fan; and high above the workmen's heads, a projecting canopy of iron was the only guide for the smoke, which was absorbed, as if by magic, in an opening in the wall close below the canopy. The anvils stood opposite the forges, and around one of these, seven stalwart smiths might be seen standing in a ring, the heavy sledge-hammer wielded by each, descending with perfect accuracy and in rhythmical time on the anchor-shaft guided by the foreman.

The foundry and the smithery are the two most interesting departments of a large factory. A separate shop is set apart for the brass-founder, and is often recognisable by the presence of dense white fumes. Fitters of every sort, rivetters of plates and boilers, joiners and carpenters, coach and waggon builders, painters and plumbers, have each their appropriate scenes of labour, or may be seen, as the press of work overflows the large area of building, pursuing their labours in the yard of the quadrangle, or even on a spare strip of land outside the gates.

The construction of railway wheels was one of the most interesting details of the work of an establishment such as we have attempted 107 to describe. Flat, wide-mouthed furnaces, in the shop set apart for this purpose, were closed by iron doors, suspended by chains that bore a counterbalancing weight, so that, at the touch of a hammer or of a pair of pincers, the door rose like a portcullis, and showed a brick hearth, covered by a sheet of flame, on which lay the pieces of iron already formed into the requisite shape, and glowing with more than a cherry-red heat, which were to form the spokes, or rather the framework that replaced that earlier contrivance. One by one, each piece of iron was removed by pincers, one end inserted in a prepared jaw, the other bent round by main force, and immediately squeezed into exact form by a lever, which compressed it between an internal and an external iron template. The triangular pieces thus formed were borne to another part of the shop, and arranged in a circle, the inner points meeting in a cast-iron box, which, when closed upon them, formed a hollow mould, corresponding to the nave of the wheel. Molten iron was then poured into this mould, and, when cool enough to be removed, a skeleton wheel, with cast nave and wrought spokes was produced. The tires, duly formed and welded, were roasting at the same time in a separate furnace, which, instead of being closed by a door, was covered by a large iron lid, like that of a gigantic pill-box. This lid, rising at a touch like the door of the spoke-oven, allowed the tires to be lifted out by a couple of men, while the uniform temperature was undisturbed, and partial cooling and consequent loss of shape was avoided. Such was the proportion of the parts, that the tire, with scarcely a touch, fitted over the outside of the skeleton prepared for its reception. The gradual cooling of the iron shrunk the whole wheel together with a uniform tension; and the bolts or rivets, that were subsequently inserted into holes drilled for the purpose, were hardly requisite to the perfection of the wheel.

The chief practical evil of the management of a large establishment of this nature, by a person who is only a magnified or glorified workman, seems to be the fact, that to persons of this description, that which is present to their senses invariably assumes disproportionate importance. It requires the discipline of mind produced by a careful and well-directed education, to give adequate care to those small and trifling details, which seem to have no voice of their own to demand attention, and yet upon which the utility of the heavier parts of the work absolutely depends. Thus a man may be strenuous in his exertions to cast a certain number of columns, or to ship a certain number of girders, by a given day. If the finance, as well as the engineering, of the establishment is under his control, the magnitude of the invoices, and the amount of the bills to be drawn on the consignees, form additional elements of stimulus. But the holding down bolts, which have to be built into a foundation long before it can be prepared to receive the iron superstructure, the nuts or the washers, which may amount in value to only a few pounds, but without which no co-ordination of the heavy work can be effected, the girder of special form, for an exceptional case, which may be required before the whole work can be got into play,—these small but vital matters are generally neglected by the bustling, bullying manager. It is thus that punctuality becomes impossible, and that contracts are made only to be broken.

CHAPTER XVI

Our own correspondent

The customers and correspondents of an establishment such as has been described in the last chapter were, at one time, surprised by an inexplicable and sudden change in the character of the letters emanating from the "Works." Instead of four or five pages of that scrambling chirography which generally betrays a disconnected dictation,—pages full of unnecessary verbiage, and obscure as to definite information,—would appear half a page in the neat handwriting of an educated man, containing not a word too much, but with not a point blinked. The new correspondent, it became apparent after a time, was not an Engineer; neither did he pretend to be one. But engineering questions, by whatsoever process, were solved when put to the new writer, and the solution was always clear, and generally satisfactory. The general grumbling, preaching, bullying, tone of the "Works' letter" had evaporated, and correspondence ceased to be a daily blister.

The author of this improvement seemed to be able to effect as much personally as by letter. A tall, slightly top-heavy, man, with an expression of countenance that was, at times, charming, generally, impenetrable, on rare occasions, flashing with sudden spite, bent over the large desk. A forehead of unusual capacity was marked by full, projecting brows, not angular or penthouse-like, but swelling with prominent curves. The manner of the new-comer was ordinarily extremely quiet; a little chuckling, almost feminine, laugh interspersing itself with easy, yet pointed, remarks. He rarely seemed to be at work, but the work contrived to do itself; at all events, it was never in arrear. For the constant scramble and drive after lost time, was substituted an air of agreeable leisure. Instead of a constant denunciation of everything and everybody, you became conscious that everything was just as it should be, or at all events, was about to be made so. The fiery manager ceased to bully, when under the fixed eye of the new conductor of the correspondence. After a while a new element appeared. Charming in society, faultless in manner, giving to the lightest anecdote a point and a pungency peculiar to his own version, this highly educated man would at times vie with Balzac himself in the colour of his stories. But, while this was the case in the society of some companions, in that of others he appeared in the character of an eminently pious man. To one associate, who soon experienced the fascination of the power of a man of the world, educated at an English university, polished by years of residence at Paris, free of all societies, and evincing at times evidence of some systematic wisdom, or peculiar style of education of thought, which could not have been acquired either at Cambridge, or in the *Pays Latin,* the aspect displayed by his new friend was that of an anxious and perplexed inquirer after truth. Tract No. 90,[65] then in the zenith of its notoriety, was to be seen on the desk. Pulling off one's hat, on coming in sight of a church, was spoken of as an essential mark of the breeding of a gentleman. Little by little many points of ecclesiastical mannerism, now known as ritualism, came to be mentioned with approval. The great desirability of a return to a Catholic unity, however this could be effected, was next brought on the *tapis.* At last, hints were given of the unusual advantages that might be secured from acquaintance, and something more than casual acquaintance, with the Directors of a silent but powerful association, who had their supporters and their friends wherever man could wish to travel. ''As to your religious views,'' said the perplexed inquirer, ''you need be under no concern. Where they are as yet unfixed or incorrect, they are sure to become clear. All that would be asked of you would be to write a letter once or twice a week, and to reply to any inquiries put in confidence; and you would thus secure the support of the most powerful organization in the world, for your future path in life.''

It happened that about the time when the veil had thus far been lifted, the two men we have referred to met at dinner, together with one or two other persons of less ingenuous confidence in human nature than the relator. The conversation was turned upon the subject of the Jesuits. ''Ah!'' said the correspondent, ''a very much misunderstood body of well-meaning priests,—very humble, but trying to be useful in their own way,—especially in educating those who have no other teachers.'' ''You speak of the priests,'' said one of the party; ''but what about the lay Jesuits?'' ''Lay Jesuits!'' said the other; ''are there such people? What may they be? I do not remember to have heard the term before.'' This little instance of serpentine wisdom, after the foregoing recommendation to make use of the support of these same lay brothers, *nomine proprio,* led to the close of the intimacy; and neither that close, nor the recollection of a *coup manqué,* was ever forgotten by the innocent admirer of the poor priests. He had one or two subsequent occasions of showing that

[65] No. 90 of the *Tracts for the times,* written by John Henry Newman and published in 1841, designed to show that the Thirty-nine Articles were not inconsistent with Roman Catholic teaching.

111

forgetfulness of injuries, even if inflicted, or only attempted to be inflicted, by himself, was not a portion of the wisdom which he had gained from the study of the Institutes of Loyola.

The exertions of this able and wily man were not all equally fruitless. One after another young men from the "Works" became attendants at a Popish cathedral newly rising in decent splendour in the neighbourhood.[66] The actual amount of proselytism that took place within a few years was considerable, but the point of chief interest in the story, is the sly and subtle character of the approaches, and the perfect adaptation of each attack to the character of the subject.

In time, this quiet and silent man seemed to grasp the control of the entire establishment. The partners might order what they liked, —nothing was done without his fiat. The course of business, or some other inducement, led him, after a time, to remove to Paris, where he acted first for the firm, and subsequently for himself. He came out as a banker. His private sources of information, the implements by which he rose to power and to the possession of the means of a luxury in which he rather too openly indulged, led to a downfall from which he would no doubt have recovered, had he maintained the original simplicity of his habits. He had sure and private intelligence of the resolution of the Emperor Alexander to conclude a peace, towards the close of the Crimean war. He purchased *rente* largely in consequence. In all political events, time, the main point for the speculator, is the least certain element. The event on which our friend had speculated was delayed. After a time it became necessary to "carry on" his speculations. To do so for another month required eight thousand pounds: he found the money, and obtained the delay. The month expired—renewal of delay was beyond his reach. His airy castle collapsed as the fatal hour struck, and he fled *incognito* from Paris, some three or four days only before the arrival of the anticipated intelligence which, so short a time before, would have enriched him by a hundred thousand pounds.

The true moral of this rapid, brilliant, and gloomily closed career is, no doubt, the illustration of the failure of brilliant talent to ensure success, unless it be chastened and directed by the presence of high and settled principle. But the observation which was most obvious to those who watched the progress of this able and unscrupulous man, step by step, was surprise that so astute a player in the game of human life should at times throw up his cards for so small an inducement, that the voyage of so well-armed a vessel should so often be imperilled for "a ha'p'orth of tar." A certain amount of self-indulgence seemed at times to be an irresistible temptation. Calmly confident in his own ability, conducting a well-planned and systematic course, bent to certain definite ends, and trained in the management of unusual intellectual powers by the subtlest of educations, the utterance of a sharp *bon mot*, however inconsistent with the professed character and design of the man, seemed at times irresistible. A degree of honest, unexpected, malevolence would flash out on occasions, the effect of which, weeks of demure and cat-like caresses would fail to obliterate. With those who in any way put themselves in the power of the corresponding manager, too freely accepted his ready kindness, or confided to him if they got into a scrape, this occasional spite would be rarely apparent. Its display was, perhaps, chiefly due to the sense of being foiled in an attempt which had to be carried on under a mask.

Self-indulgence, in the ordinary sense of the term, appeared to be altogether foreign to the habits of the scholarly man of business. Late and early his pen was ever ready, his attention to any call of duty prompt, his resource unfailing. But perhaps the strongest claim to the admiration of a cultivated mind, which he tacitly urged, was that of a heroic and elegant poverty. He seemed to do without money—to require none, to care for none. All that he had avowedly to live upon, was a salary less than the weekly wages of a foreman of smiths. His house was, undisguisedly, a labourer's cottage. Yet in that cottage was to be found a spare bedroom, if little more than a closet, for a friend, and a simple and graceful hospitality, which converted the transformed hovel into the house of a well-bred gentleman. Under the cloudy canopy of eternal smoke, that little home was clean and bright. Articles of furniture were few, but a special elegance, or a special history, distinguished each; and, besides, there was little room to spare. The chief ornament of the cottage was its mistress, an extremely pretty little woman, with a glorious coronet of hair. Very young (she made her own mother a grandmother at the age of thirty-two), she looked rather an elegant adornment to her husband's home, than a housewife bearing the load that presses on a poor man's wife. None of the toil, none of the detail, of housekeeping seemed to fall upon her. Everything was thought of, arranged, executed, by her husband, and that in such a way, that there seemed to be no thought, no care, no exertion. Everything, at the right moment, seemed to have provided for itself.

The difficulty was, to leave the man's society, when he had made up his mind to be agreeable. Nothing seemed to arrest his flow of chat, of anecdote, of fun. Night brought no fatigue—the same quiet readiness, the same sparkling reply was always forthcoming. If good company passed the bounds of good fellowship, it appeared neither to quicken his pulse, nor to flush his cheek. Night or day, toil or chat, abstinence or full conviviality, seemed all alike to him. ''How is it,'' said an acquaintance, ''that wine affects you no more than water? Would no quantity overcome you?" "I suppose," was the reply, "that something unreasonable—drinking off a bottle of brandy, for instance—would produce a sudden collapse—that I should become suddenly insensible, and sleep off the effect. But I do not think my brain can be affected in any other way." It would have been well if this grand state of physical discipline had been constantly maintained. It did not last for more than a year or two. As his power became established, and his services indispensible, this rare sobriety of character evaporated. Not only so, but it was replaced by rather too strong a dose of the opposite quality. Habits are not thought to be inconsistent with the conduct of business in Paris, that would be scandalous in London, and simply impossible in a country manufacturing district. What would an English Engineer say, to having a rendezvous given him at 11 or 12 a.m. in a coffee-house, and then finding his correspondent finishing a champagne breakfast with a game of dominoes? Nothing is more common in Paris than such an appointment. Englishmen who adopt foreign habits very often go far beyond the foreigners whom they imitate, but the worst of the French hours of business is the frittering away of time which it causes. On the other hand, many of these late breakfast-eaters have been steadily at work for four or five hours, commencing in the early morning. But, in the case of our friend, the debauch was more serious, and the effect soon became visible. Champagne and turtle lunch, dinners at the Hotel Bristol, the distraction of more than one domestic establishment, interfered with the steady working of the intellectual machinery. The tall and slender form, graceful in its aspect (except for a certain effect produced when walking, which gave the impression that the legs were tied together at the knees), became corpulent and heavy, without losing the last-named peculiarity. The fine lines of the face filled up, the cheeks grew fat and puffy, the balanced manner and calm, measured speech gave way to frequent expressions of irritation. You saw before you a man, who seemed to feel that he could, very often, and for a long time together, lay aside a mask which he had long worn with pain. He kept it still at hand, but only to put on when necessary. Nay, those who surprised him without it, were no longer deceived when it was assumed.

A man a few years older than himself, his inferior every way in education and ability, his equal in integrity of principle, but his superior, to some extent, in the power of keeping up an acted part, came at one time into close business relations with the firm, which was still, to a great extent, controlled and manipulated by the subject of our sketch. The two became natural enemies, yet the points they had in common formed a sort of masonic bond between them. The elder, not aware of what was the hold of the younger on those whose nominal servant he was, laid a pleasant little trap, into whch the other fell with the greater readiness, inasmuch as he was careless as to the result. A gay and lively dinner, with abundance of excellent wine, was carried on with such hearty good fellowship, that our hero, whose nerve was already shaken, became palpably inebriated. The other kindly put him into a cab, and drove him to the residence and office of one of the partners, where business of some importance had involved an evening *sederunt*.[67] The air and the brisk drive supple- mented the generous liquor, so that the assistance which the host ostentatiously volunteered, in order to guide the other from the cab, and into the presence of his principals, was actually necessary for steadying his gyratory movements. He looked round with an angry glare, made one or two efforts to discourse, and then, with an im- portant abruptness, ''Cantshtay—musrite letto Roschild 'twonce,'' made off to his own premises. The well-intended malice produced no apparent effect.

[67]i.e. session.

How far this gentleman was, or was not, connected with any organ- ized operations of the Society of Jesus, the relator might have been better able to judge but for the *contretemps* before described. Probably

those who could answer the question would anticipate inconvenience from so doing. The French novelist who tells us of the proceedings of the *Père d' Aigrigni,* and the more terrible *Rodin,*[68] is for the most part, in this country, regarded as a romancer. It is, of course, possible that the affiliation of the letter-writer to his mysterious association was imaginary or fictitious—a claim invented to surround his character with an effective mystery. On the other hand, a subsequent study of all that can be learned by the uninitiated of the ''Constitutions'' of Loyola, led to the conclusion that the peculiar ability of the man had been formed, or greatly aided, by a mode of study unknown in any Protestant college, or establishment for education. The intimate relation that existed between him and the higher Romish clergy of the neighbourhood, was as well known as was his success in making proselytes. These relations, and this success, followed too close upon the time when he represented himself as an Anglican in doubt, to allow of the possibility of that statement having any foundation in truth. The gradual approach made to the question of affiliation, and the sudden dropping of the subject when success became improbable, were remarkable. Again, there were little circumstances which, from time to time, excited curiosity. When the account commences, the letter-writer was receiving only the small salary of £150 per annum, on which he was supporting his wife, with an annually increasing number of olive branches. His private resources had, according to his own account, been dissipated. But every morning saw the *Times,* then a costly luxury, laid on his desk, from the office of publication. Occasionally, these papers would remain unopened for days together; but that the best means of information, and that at a sensible cost, disproportioned to his apparent position, were placed at our friend's disposal, there can be no doubt.

A classical and mathematical scholar, a master of modern languages, a man possessed of that steady glance which seems to have power over violent and hasty men—something like that of a keeper over a maniac—but who rarely attempted to use this power, as he usually turned his glance aside when you looked him in the face, though his eye was rivetted on you at other times—this man resembled a costly and highly-finished instrument in which there was some lurking imperfection. His power to read and to control the minds of others, was wasted for the sake of worthless and momentary triumphs. His elegant frugality, his sober self-control, his perfect readiness for efficient action at a moment's notice, were discarded for a Sybaritic luxury, which might have been natural in a mere vulgar man, when raised to the command of wealth, but which surprised those who knew the earlier part of his business career. His unquestionable abilities led to a disastrous *fiasco.* His very sources of power proved the causes of his downfall. His character, or at all events his ecclesiastical relations, were enigmatic, and the enigma is one, of which the full solution could not fail to be eminently instructive.

CHAPTER XVII

Construction of the broad-gauge lines

When the broad-gauge lines first began to push forth their feelers, beyond the limits of the district traversed by the original trunk, the sentries and out-skirmishers of the narrow-gauge system became aware of the presence of a school of construction very widely differing from their own. Some of the unnecessarily original peculiarities of the Great Western Engineer were abandoned as his experience increased;[69] but the impression which was produced upon the Engineers and contractors who had been formed under the command of Robert Stephenson, was not at first favourable to the very different method according to which things were managed by Mr. Brunel.

The first peculiarity, and that which underlay and augmented every other, was the personal share which the untiring energy of the latter Engineer led him to take in every detail for which he was ultimately responsible. The staff of Mr. Stephenson, although it cannot be said to have had a military organization, yet to some extent resembled an army corps in division and subordination of duty. Each officer was the servant of the Company. Each had his own limit of function and responsibility; and although Mr. Stephenson well knew how to show, from time to time, that not the smallest detail would escape his attention, if it involved what was wrong, the order of his office was not such as to overburden the Engineer-in-Chief with details that fell properly within the competence of the Residents, or even of the Subs. But the Engineers on the broad-gauge lines appeared to regard themselves less as the officers of the Company than as the channel of the will of Mr. Brunel. His Residents no more ventured to act without his direct authorization, *ad hoc,* than did any inspector on the line. Of course there would be differences due to the greater or less distance from town, or the greater or less personal acquaintance with the Chief, possessed by his former pupils, or by any others who held office under him; but, to a very late period, the personal knowledge and direction of the Engineer-in-Chief was the secret of the pervading energy that directed all his great and costly works.

With this method of administration, resembling that of the Spanish

[69]Notably his method of laying the broad-gauge track on piles.

118

monarchy rather than that of a constitutional government, were blended two other peculiarities that at first assumed an extreme rigidity of character. One was the scientific knowledge of a pupil of the "École Polytechnique,"[70] which contrasted rather sharply with the general and time-honoured English system of the "rule of thumb." The other was an extreme and unprecedented insistence on excellence of work. These two features of Mr. Brunel's system were indeed intimately connected with one another. A smaller quantity, for instance, of accurately bedded and highly-finished masonry, might afford as much security to a structure, as a larger quantity of rougher work. The cohesive force of the best hydraulic lime was an element in the calculation by which the quantities of Mr. Brunel's bridges were kept below those of his rivals and contemporaries. To persons familiar with French engineering, this was easily intelligible; but to English contractors it was not so. A man made a tender for what was called a rubble bridge. He saw, if he gave proper attention, that the specification was unusually strict. But he did not, until taught by experience, anticipate that there would be an inspector on the ground, who was only an embodied and very quick-sighted specification. He found, before he had done, that Mr. Brunel's "rubble work" meant, unusually good ashlar in small courses. He found that he had given prices per yard, when he ought to have estimated them per foot. At the present distance of more than thirty years, a sum of £120,000 has been charged against the expenditure of the Great Western Railway, between December 1865 and June 1867, which represents the difference of opinion between the Contractor Macintosh and the Engineer Brunel, as to the powers of the latter, and the legal force of his specification.[71]

Contractors took the alarm. Here one was told of an order to remove instantly from the works a large number of bricks, as being inferior to specification; the apparent cause being that, for the sake of his own credit, the builder in question had made use of a very superior quality of bricks for the face of the work. The Sub-Engineer inexorably pointed to the clause, that all the bricks used on the work should be of the same quality, and demanded that abutments, and even footings, should be constructed throughout of the best malm face bricks. In another case, a contractor who had built a river bridge of highly ornamental rustic-work, under the daily inspection and constant approval of one of Mr. Brunel's Residents, was ordered to smooth-tool all the faces of the two piers, because there was less than the specified width of water-way at high tide, between the extreme projections left within the channels of the rustic-work.

Sometimes the Residents would exceed their power or misread their brief, and come to grief in consequence. There was an amusing

[70]I. K. Brunel was never a pupil of the Ecole Polytechnique. His French education began at the College of Caen, went on at the Lycée Henri Quatre in Paris, and ended in apprenticeship to Abraham Louis Bréguet, one of the greatest of all watchmakers.

[71]See note 58.

instance of this in the case of one of those who was always in a state of incipient trembling under the eye of the Engineer-in-Chief, but who solaced himself for the discomfort by more than necessary harshness to those under his own orders. As he was, moreover, a man of hesitating and uncertain manner, of a memory so unreliable as to throw much doubt on his statements, and of an education which, whatever it had been, had not produced the virtues either of military exactitude, or of very evident acquaintance with mechanical principles, he was not a favourite either with his staff, or with his natural enemies, the contractors. One of the latter, a man of ample means, had taken several large contracts under the idea of readily increasing his income. The crafty and experienced foreman who had induced him to enter into this speculation, and on whose knowledge and ability the capitalist had altogether relied, became so much elated with his own exaltation to the command of a large number of men, that he considered sobriety to be thenceforth superfluous. When he had spent some weeks in continuous intoxication, it became necessary to put him aside—not an easy operation. The arrangements of the contractor naturally fell into extreme confusion, a state which was augmented to the utmost by an almost constant downfall of rain for some ten months, with hardly a day's intermission. To understand the full force of the latter misfortune, it must be told, that the county, where the works were situated, had a soil of very tenacious clay. The contractor was obliged to have recourse to a young Civil Engineer to help him out of his sorrow, to finish the works, to draw up the accounts, and to settle them finally with the dreaded Engineer-in-Chief. The state of active and acrid hostility, into which the Resident Engineer' office had been lashed by the numerous shortcomings of the inexperienced amateur, was not one of the least formidable difficulties of the situation.

On one occasion, this Resident Engineer requested the younger man to call early at the office on a matter of urgency. The other duly appeared. "Are you aware," said the official, "that your slopes are wrong?" "No," was the reply. "They are indeed; I have been measuring myself. I have measured the battering rules by which they are set out, and they are quite wrong." The other was taken aback. "But," said he, "the *line* is right." (For it had been just prepared for the permanent way). "I find no fault with the line," said the Resident, "but the slopes are all wrong—I mean those that have just been turfed and sown." "It is a pity that you did not point it out before that expense was incurred," said the other. "That was not my business," replied the Resident; "the contractor is responsible." "Well, but—" said his opponent, beginning to recollect himself—"the *top* of the slope is regulated by the fencing, and the fencing was set out by your own pegs.'' ''But you know that the specification especially states that the contractor is responsible for setting out the works. If we have been considerate enough to save you that trouble, you cannot turn round on us and say that we did it wrong,'' said the Resident. ''Then what do you wish me to do?'' ''You must remove all the soil, reduce the slopes to the proper inclination, and soil and sow them over again.'' ''Why, that will cost four or five hundred pounds!'' ''I have nothing to do with that.'' ''Well,''said the other, ''it is rather too bad to have to do, over again, work carried on and completed under your eyes, and to your pegs. The schedule price for it is high; suppose we split the difference, and I charge you half-price for doing it. The moral responsibility lies at your own door at all events.'' ''You will find nothing about moral responsibility in the specification,'' quoth the Resident.

The Engineer of the contract made a rapid retreat, in order to control his temper. Galloping to the site of the principal cutting, he called out for the production of the batter rules, or bevels, to which the slopes had been ordered. On their appearance, a fact which had escaped his memory became at once apparent. On each of these implements, branded in by *his own hands* with a hot iron, was the ratio of the slope, and the number of the cutting to which it was applicable. To apply the two-foot rule to the sides, was the work of half a minute. They were accurately true. The bevels being then applied to the surface of the condemned slope, the same result followed. *Everything was correct.* The sole revenge which the Engineer took, was to write to the Company's Resident: ''Dear Sir,—Referring to our conversation of this morning's date, you will find, on further examination, that you must have measured one containing side, and the hypothenuse, of the batter rule, instead of the two containing sides, as both the rules and the slopes are perfectly accurate.'' The subject was not again referred to by the Resident Engineer.

The inspectors on the broad-gauge lines were frequently men whose method of discharging their duties would not have recommended them to the confidence of the Engineer-in-Chief, had he known all. Their general *rôle* was to exaggerate as far as possible the minute requisitions of the specification, and to render the operations of the contractor as expensive, as difficult, and as unsatisfactory as possible. The ingenuity devoted to this end was extreme. But some of the sternest and severest of these men had a certain elasticity of conscience. It become obvious that, under certain conditions, their opinions were liable to change. In a word, and not to speak jestingly of a subject of great importance in connection with public works, it is certain that an attempt to enforce an altogether unusual degree of strictness, in the execution of large undertakings, offers great temptations to bribery. The contractor, who sees that more is continually demanded of him than he ever contemplated, and more than he believes to be of any practical advantage, is induced to wink at methods of removing a source of costly and unreasonable vexation, as it appears to him, by incurring an expense which is, no doubt, unjustifiable, but which he may be easily led to regard as the only way of avoiding ruin. Over-strictness is sometimes as demoralising as over-laxity.

Again, the reputation which the earliest contracts on the Great Western acquired for the Engineering management, must have raised very materially the schedules of prices of a large proportion of Mr. Brunel's contracts. Contractors declined to compete. In large districts of country the broad-gauge contracts at one time actually went a-begging. Some fell into incompetent hands; some were put into the hands of very competent persons at extravagant prices. No small proportion of the heavy cost of the works executed by Mr. Brunel, has been owing to the fears entertained, and justly entertained, by contractors, of the power entrusted to the inspectors to render the work ruinously expensive.

While great attention was given, from the very first, to the specifications of Mr. Brunel's contracts, there was a marked indisposition, by no means peculiar to this Engineer, to supply to the contractors that detailed scientific information, at which they have so much less facility of arriving by their own calculations, than has the designer of the works to be let. In all the early railway contracts, a responsibility as to quantities was thrown upon the contractors, which it would have been more economical, as well as more dignified, for the Engineer-in-Chief to have frankly assumed. The idea at first entertained of the best method of letting a contract was this:—The line was accurately surveyed and mapped, levelled and sectioned, having been carefully set out with pegs; coloured drawings were made of every individual work of any magnitude, together with sheets of general drawings of fencing, culverts, gates and crossings, and permanent way. A complete, signed, set of these drawings was left for an advertised time at the office of the Engineer, for the inspection of contractors, who were allowed to make tracings and copies. A specification was prepared, in accordance with the drawings, and the contractor was expected to tender to execute the whole of the works comprised in the length of the section, as detailed in the drawings and specification, for a lump sum. He was also to fill up two schedules of prices, one according to which the work, measured up from month to month, was to be paid for on account, and one according to which any additions or deductions ordered by the Engineer-in-Chief in writing were to be allowed for. Later in the day a single schedule was used for both purposes. It was customary to mention the quantity of earthwork which the section was estimated to show, but this was done with the express provision that the Engineer was not responsible for the accuracy of the estimate, but that the contractor was to take the entire risk in consideration of the stipulated sum.

Practice soon made it apparent that increased responsibility meant increased cost. A detailed estimate was necessary for the guidance of the Engineer-in-Chief, and to enable him to advise the Board as to the merit of competing tenders. A minute and accurate calculation of the contents of this section, distribution of the quantities, arrangement of the leads, and admeasurement of various masonry works, was essential, sooner or later, to the Engineer-in-Chief. It was far better, in all respects, for this to be done in the first instance. Where this was the case, there could be no good reason to deny to the various persons about to tender, the facility of a communication of the results of a purely scientific work. Much time was thus gained to the Company, expense was saved to the contractors, blunders were avoided, and the reduction of unnecessary risk was such, as to lessen the average amount of tenders by at least five per cent., and often by much more. It is hard to see that any undue or disadvantageous responsibility could be thrown on the Engineer, by having all his work fairly placed on the table, instead of calling for comparatively uneducated men to make guesses more or less in the dark. Still, it was but slowly that so obvious a reform was effected. The plan which has been adopted so recently as during the construction of the London, Chatham, and Dover Railway[72] is the opposite of the original lump-sum form of contract, consisting merely in a schedule of prices, without any stipulation of quantities, the latter being left to the determination of actual measurement. The objections to this course, which possesses the *prima facie* advantage that no charge can be made except as payment for a definite quantity of work actually done, are, that it leads both company and contractor to bind themselves to arrangements with the actual magnitude of which they are unacquainted, and that it presents the constant temptation to the latter, to swell his account by the performance of an unnecessary quantity of work, an evil of which there is abundant proof on the Kentish line to which we have referred.

In providing for letting the contracts on the Manchester and Birmingham Railway, a line very ably constructed by Mr. Buck,[73] one of the earliest Residents of Mr. Robert Stephenson, the just mean was very happily attained. Drawings and specifications were prepared in the usual detail. In addition to this, printed lists of quantities, giving earthwork, turfing of slopes, fencing, lead and distribution of materials, contents of each bridge or work in masonry, number of gates and level crossings, length of line to be laid and maintained,

[72]Here the engineer was Thomas Russell Crampton (1816-88).

[73]Buck is described on p. 11.

and any other items which the contractor was to include in his lump sum, were furnished to all those who were invited to tender. The result was, that the contracts were freely competed for by respectable men, and were taken under the estimate of the Engineer.

The manner in which the attention of Mr. Brunel's staff was concentrated on the technical minuteness of the specification, and the unexampled finish of the work, produced an undesirable effect on the estimates. Very many of the details of the work were left to the discretion of the Engineer, and thus came under the schedule of prices for extra work. Thus on the settlement of some contract accounts which had to pass through the office of the Resident Engineer above quoted, he was heard to remark that the amount of extras, which was fifty per cent. on the contract sum, was "about the regular thing." Alterations of design entered largely into the sources of cost, compensation having often to be made for abandoned work. But experience was never thrown away on Mr. Brunel, and while his earliest works left the question of ultimate cost very much to come out as it would, nothing could exceed the accuracy and the detail of the scientific information which was prepared before the putting in hand of his later contracts. In country of at all a sidelong and hilly description, cross sections were taken and plotted at every chain, or even more frequently, so that the direction of the line might, if necessary, be modified in the office, with perfect certainty as to the effect of such alteration in increasing or diminishing the quantities of excavation and embankment. In fact the contract bill, as it would ultimately have to be rendered, unless any written orders from the Engineer-in-Chief should modify any of the provisions of the specification, was prepared in Duke Street,[74] and served as the basis of the whole execution, and fortnightly admeasurement of the works. The regularity and comfort thus introduced was also an element of economy to the companies that had to pay the contractor.

[74]Brunel's offices were in Duke Street, Westminster.

124

CHAPTER XVIII

The author of the broad gauge

The preceding references to some of the disadvantages which attended the earlier contract work of Isambard Kingdom Brunel must not be held, either directly or indirectly, to detract from the well-established fame of a man who, in the opinion of every competent judge, was second to none in the profession of which he was so distinguished an ornament, and who possessed many rare and noble qualities in which very few could claim to rank even as second to him. The speed and luxury of English travelling, both by land and by sea, is probably due even more directly to the genius of Mr. Brunel, than to the labours of any other man. It was the rapidity attained by the giant engines on the broad gauge, that drove the engineers of the rival lines to substitute long six-wheeled engines for the jumping four-wheelers that were first placed on the London and Birmingham.[75] The Engineer of the Great Western Railway was also the designer of the Great Western steamer; as well as of that later and fatal triumph of science, the Great Eastern. Those who had the privilege of a sufficiently close access to Mr. Brunel to enable them to trace and watch the workings of his mind, must have been very incompetent judges if they did not become aware of the presence of a genius of the highest order. Modern times have seen no more striking proof of the theory of Aristotle, that virtues may fail by excess as much as by defect, or, in other words, that the excess of any virtue is a vice. The imperfections in the character of Mr. Brunel were of the heroic order. He would have been, commercially speaking, a more successful Engineer, had he possessed a less original and fertile genius. His untiring and insatiable industry wore out his iron frame prematurely, as to years, though late, if measured by accomplished work. His conscientious resolve to see with his own eyes, and to order with his own lips, every item of detail entrusted to his responsibility, brought on him an enormous amount of labour, which, on another and a more easy system, would have been borne by subordinates, perhaps with equal advantage to the public. His exquisite taste, his perfect knowledge what good work should be, and his resolve that

[75]This is an over-simplification. It is true that the broad-gauge Great Western Railway never had four-wheeled locomotives; but its first successful machines were the six-wheeled engines supplied by Robert Stephenson & Co. in 1837 to a design evolved for the service of the standard gauge four years earlier. See J. G. H. Warren, *A century of locomotive building*, 1923, 79, 340.

125

his works should be no way short of the best, led rather to the increased cost, than to the augmented durability, of much that he designed and carried out.[76] His boundless fertility of invention, and his refusal to be content with what was good, if he saw beyond it what was better, led often to disproportionate outlay. Devoting to the scientific mastery of his profession the years during which some of his most successful rivals and friends had become acquainted with the details of actual work, and with the habits of the workmen, he stood at a disadvantage when he first assumed, at an early age, the unchecked control of large public works. It was not peculiar to Mr. Brunel to be educated at the expense of the shareholder, and by the progress and the difficulties of his undertaking. But the difference between him and those who were nearest to him in ability lay here, that he saw always clearly before him the THING to be done, and the way to do it, although he might be deficient in the experience to direct him in the choice and the management of the human agency, which was necessary to effect his designs. Mr. Stephenson, on the other hand, knew how to derive from his staff and his friends a support and an aid that carried him, at times, over real engineering difficulties, with a flowing sheet. Both wrought for fame; both wrought for the benefit of their clients. It may be, that the perfection and success of each individual work was more the study and aim of Mr. Brunel; the return of benefit to the shareholder the more present idea of Mr. Stephenson. Mr. Brunel, when asked if he would build a steam-vessel a thousand feet long, said, nothing would please him better, but that, to meet ignorant prejudice, he would recommend building *a little one* of 750 feet long to begin with. Mr. Stephenson said he looked forward to the time when no poor man should be able to afford to walk from his home to his occupation. The former prepared the luxurious cabins of the Atlantic steamers, and the commodious sofas of the broad-gauge carriages; the latter opened the way for parliamentary and workmen's trains.

There was another part of the character of Mr. Brunel that endeared him to those who knew him, and that should be no less commemorated, to his honour, than the design of those noble works by which his genius reared an imperishable monument to his memory. It was this, that with advancing years and increasing power his character mellowed. With most of his contemporaries, youth was the happiest time, at least as far as regarded their intercourse with their subordinates. With Brunel it was, until his health began to fail, more pleasant to have relations the older he grew. It is probable that the high sense which he entertained of responsibility, and the eager and noble thirst for renown, impressed on his earliest dealings with English contractors a harsh and arbitrary character, which he was able to

[76]Again an over-simplification. Brunel went to extraordinary pains to design timber bridges for all the railways he had laid out, and a good many of those —notably the series on the Cornwall and West Cornwall Railways—took this form partly in order to keep the first capital expenditure down. See the very interesting chapter on his use of timber by L. G. Booth in *The works of Isambard Kingdom Brunel,* ed. Sir A. Pugsley, 1976.

lay aside when he found his power to be unquestioned. He was highly impatient of insubordination, or of contradiction. Very much more was to be done with him by following his course, than by attempting to take one's own.

The final settlement of the account of the contracts on which the little difficulty as to slopes, previously mentioned, occurred, may be described, as illustrating very fairly the character of Mr. Brunel, at a period when he had attained that unquestioned authority in his profession to which, for so many years, the Directors of his various enterprises were tolerably submissive.

After a considerable amount of labour, the amounts in question had to be made up, and were duly presented to the Resident Engineer. That gentleman, looking, not at the items, but at the total, rejected them in block, and was so far from possessing either the patience, or perhaps the power, to discuss them, that the whole matter was referred to the personal decision of the Engineer-in-Chief. He made an appointment with the framer of the accounts to meet him after a half-yearly public meeting at Gloucester. He had left London by the mail on the preceding evening, and from that time to about five o'clock on the day appointed, a continued round of appointments had left him no time for sleep, except such as he might snatch in his carriage between the several points of rendezvous. He was rapidly disposing of a mutton chop, when the younger man entered with an enormous bundle of papers. "I hope, So-and-so," said Mr. Brunel, "you are not going to ask me to look at all these papers." "I hope not, Sir," was the reply. "I hope that you will be satisfied with two or three that I have drawn up expressly for your approval; but if any difficulties arise, I am provided with my proofs and vouchers, 127 and I am sure you will rather read them every one than fail to do my client justice."

Happily it was not necessary to inflict this labour on the Chief. The bearing of the specification on one or two points, as contended for on the part of the contractor, and as opposed by the Resident Engineer, was admitted by the framer of the document. This point settled, the extra charges, following one by one, were successively admitted. At length came one which elicited the remark, "I do not see your ground for that." "No," said the other, "it is a weak point, and one which I should hardly have brought forward if I had anticipated the very impartial way in which all the preceding items have been dealt with. I withdraw it." Out of the seven or eight thousand pounds which had been summarily disallowed by the Resident Engineer, there now only remained one item, of about £800, which arose thus. The bridges on the contract were specified to be of stone, but power was reserved by the Engineer to order them of brick, at an increased price. In the unusually wet season which supervened, it became all but physically impossible to convey stone from the quarries to the site of some of the bridges, to which no roads were to be found, for a tangle of ruts full of tenacious wet clay, on which the bottoms of the carts rested, while they were pushed forward like sledges, with their wheels cleaving through, not revolving in, the mud, had occupied the site of the former roads. The Engineer's office had been applied to, for permission to substitute brickwork, on the ground of expedition; and the substitution had been duly ordered, the question of price having been mooted by neither party. On behalf of the contractor, it was therefore urged, that the plain terms of the contract entitled him to the price for brickwork, the Company having the value in a more durable material than oolite rubble. The Resident Engineer replied, that the permission to use brick had been given at the request of the contractor, to save expense in carting stone, and that he was entitled to no extra charge on that account. Mr. Brunel held to the view of the Resident. The other, finding that the opinion was tolerably firm, and unwilling to give up an item for which there was much to be urged, proposed to refer the point to a third Engineer, quoting a clause in the specification to the effect, that in case of a dispute as to accounts the contractor should be at liberty to call in an Engineer, who should confer with the Engineer-in-Chief; and that, if they did not agree, they should appoint an umpire, whose decision should be final. "I think we are quite within the corners of the contract," said the claimant. "On this point the Engineer consulted by the contractor, cannot agree with the Engineer of the line, and I must ask you to name an umpire." Mr. Brunel drew himself up. "In that case we must consider this conversation

128

as not having taken place. I have met you as a gentleman, and, I think, liberally; but if you talk of putting an umpire over me, you must make out your whole case as against me!'' That of course was conclusive. ''I thought it right to make the best fight I could for my client,'' was the reply; ''but if you put it in that way, I have no more to say. I really think there is a fair claim, but I must leave it altogether to your sense of justice. I am much obliged to you for the patience and kindness with which you have met me.'' He put up his papers, and as he was taking leave: ''I'll tell you what, So-and-so,'' said Mr. Brunel, ''if you think you really ought to have that eight hundred pounds, I will not object to giving it to you, if you will obtain the assent of your old friend Captain Moorsom, who may possibly have to take over this part of the line for his Company.'' The sum was paid.

It is impossible to avoid taking comfort at times from the admitted weaknesses of great men. Some two years after the date of the last conversation, the relator was accompanying Mr. Brunel over a portion of the South Wales Railway. They passed through the little town of Kidwelly, where the sturdy round tower of the castle still recalls the aspect of feudal times. ''Do you know,'' said the great Engineer, ''I always feel very sneaking when I pass this place. When we first obtained the Bill for the line, the good people here were overjoyed. They held a public meeting, and resolved to present me with the freedom of the town. It was drawn out in form, and sent me with a letter from the mayor, and, in fact, in a gold-plate box. Of course I felt gratified, and I waited a day or two before acknowledging the letter, as I was extremely busy, and I wished to do so in suitable terms. So a week or two slipped by, and I thought that I must make my apologies in person. But from that time to this the opportunity has not presented itself. I am ashamed to say that I have never even thanked them, and I feel as if I must rush through the town whenever I have to pass it, from being unable to excuse my neglect.''

It has been always regarded as a subject of the utmost interest, to ascertain the method of thought natural to those who have left their marks deeply scored in the history of mankind. Thus it is recorded that the method of Sir Isaac Newton was, to keep the subject he had under investigation continually before his mind. After a time a bright light seemed to dawn on it, as if from some source of illumination. It was on the occasion of this Kidwelly conversation, or at all events about the same period, that a slip had taken place in some marine works at Neyland, which Mr. Brunel had been down to examine. After looking at the spot with minute attention, waiting for the ebb of the tide, probing the sand, and taking every possible means to ascertain the actual facts, Mr. Brunel left the spot, and maintained silence for a considerable time. At last he spoke like a man waking out of sleep, gave a few plain and precise directions, and ordered a minute report to be preserved of the mode in which they were carried out. He was asked, not as a matter of professional direction, but as a question of intellectual interest, how he had come to the conclusion. "All the time that we have been travelling," he replied, "I have been trying to fancy myself under the sand, and at the foundation of that wall. I have been trying to realize the scene, and to make up my mind as to what was actually taking place down there. At last I seemed to see it plain; it was easy then to order what to do, and now you will see whether I am right or wrong." It will not be doubted that the insight thus obtained was mechanically true, and that the method employed was successful.

With this constant absorption in his profession, and with the clear insight obtained, alike into those physical relations which are the province of the Engineer, and those outlines of human conduct and human action, the appreciation of which has so much to do with the success of parliamentary practice, or the upshot of an examination before a Committee, Mr. Brunel allowed himself the relaxation of quaint and original speculation on matters disconnected with his duty. He was almost the only man of taste, and of practical ingenuity, who was ever heard to defend the claims of the chimney-pot hat on the submission of the Englishman. In his own office, when the inroads made on his health by his constant disregard of the claims of nature to repose, began to manifest themselves by cruel neuralgic pain in his over-wrought brain, he wore a black velvet cap; but on all other occasions, in the works, in the train, even in his own convenient travelling carriage, he wore the chimney-pot hat. One of his reasons for approving it was, it gives warning in cases of danger, as it must be smashed before the head can be struck. "Then," he said, "it is at once warm and airy, and you cannot improve upon it." Questions which excited much popular attention, as the origin of mankind, and the earlier state, geologically and palaeontologically speaking, of the world, were dismissed by him with an unexpected, and summary question, "Why should not everything always have gone on as it does now," said he. "That is the simplest way of looking at it; and I am rather of that opinion. I do not see why there ever should have been a beginning."

For the construction of the Great Eastern steamer, England paid the highest price possible to be demanded in one instalment. Coming as it did after a long course of unsparing overwork, the anxiety attending that gigantic vessel, which in some of its details almost attained the quality of organic structure, no doubt led him to the grave. He pondered over every detail of the young giant, with the affection of a parent. He had been requested to build it in Milford Haven, where a commodious creek, acting as a perfect natural dock, would have enabled him to construct it afloat, and thus to avoid the peril and evil of the launch. He pondered over the proposal long, and only declined it on account of the distance from London, which would interfere with the personal attention which he intended to devote to every item of detail. For himself, for his profession, and for the country, the decision was highly to be regretted. It is, perhaps, not sufficiently borne in mind, that the great misfortune of the launch, an engineering misfortune such as was without parallel in the career of Mr. Brunel, was due entirely to his anxious care for human life and safety. Had his arrangements been carried out, the giant structure would have floated in the Thames, within a few minutes from his giving the signal. But the needless presence of the crowds of spectators invaded the space which he had desired to be reserved for the sway, not only of the moving monster, but of the wave which it would drive across the Thames. Boats full of foolhardy trespassers were tempting submergence. It was to protect the occupants of these boats from the effects of their own contempt of danger, that the movement of the launch, when well under way, was suddenly checked by Mr. Brunel. No substructure could resist such a strain. The whole weight of the leviathan acted as a sledge-hammer to drive down the ways and supports, and the long, tedious, and costly work of the launch, was the unavoidable result of that act of kindly commiseration.

Those who never saw Mr. Brunel under examination, or rather under cross-examination, neither knew the powers of the man, nor knew how the ablest Counsel may be foiled, with their own weapons, by one who knows both what they mean, and what he means himself. On one occasion the Examination-in-Chief had not produced a very satisfactory impression on a Committee of the House of Commons, as regarded the scheme supported by Mr. Brunel. The opposing Counsel girded on his armour for the cross-examination. The witness sat quietly—a stranger would have said stupidly—awaiting his attack. It came in earnest—the keen attack of an able advocate. Question after question did he produce from his repertory of offence. Very brief, dry, direct, answers were given to each question. At last the Counsel found that he had so proved the case of his opponents, by the evidence which he had thus unintentionally elicited, that he sat down with a much less triumphant air than had marked the commencement of the struggle. The decision of the Committee was easily to be foreseen. "I say, Browbeat!" said Brunel, as they cleared the room, "I should recommend your Directors to retain you to cross-examine me again next time we go before Committee."

One other anecdote of the dry fun and prompt resource, of the man. The Quaker Director, who was referred to in an earlier chapter, gave a great breakfast to celebrate the opening of one of the lines with which Mr. Brunel was connected. Everything was there that could be wished for, or thought of, with the sole but important exception of wine. Champagne was an abomination to the Friend. He was a severe teetotaller. He was not even content to throw the onus on the conscience of his guests, as was once done by the no less plain-spoken lady, who presided over a dinner table at Birmingham, at which a person who has supplied much of the anecdote of these pages was a guest. "There is wine on the sideboard," she remarked during an ugly pause; "we think it sinful to drink it; but it is here for those who have not thought it essential to abstain." But Brunel could not make a breakfast on pines, grapes, and coffee. He asked for a pint of beer. The host was inexorable. "I cannot breakfast at this hour without something of the kind," said the guest. "I must help myself, and return to finish your good things." Followed by two or three of the party, he left the Quaker's banquet, and, entering a public-house hard by, enjoyed a good draught of ale, returning contented and in perfect good temper to his intolerant entertainer.

132

CHAPTER XIX

The Engineer of the London and Birmingham

The personal acquaintance with Mr. Robert Stephenson possessed by the writer of these pages, although earlier in its commencement, was far less intimate than that enjoyed with his great rival. It is not within the scope of the present work to give anything at second-hand, or to relate characteristic anecdotes that may be found elsewhere. Mr. Stephenson is worthily entombed in Westminster Abbey. His fame is a part of the public property of the English people. On his faults, there is no occasion now to turn the bull's-eye light of criticism. But one instance of gentlemanly and honourable feeling may be recorded, which is unknown to his biographers.

A gentleman, who had been a pupil on the London and Birmingham Railway, had occasion to visit Hâvre-de-Grace shortly after the commencement of the Paris and Rouen Railway. The Syndic and principal merchants of Hâvre had become aware of the importance of connecting their port with the capital, and regarded the junction of Paris with Rouen, rather as the establishment of the latter city as a rival port, than as a necessary link in their own future chain of communication. It is true that the difficult navigation of the Seine is such as to give, to a great extent, a monopoly of commerce to the port, as the *embouchure* of that sand-encumbered stream. The lower portion of the course of the Seine is the scene of one of the most remarkable physical phenomena of European hydrography—a phenomenon which, on a reduced scale, occurs in the Severn. The mouth of the great Norman river presents, on the chart, the outline of a gigantic funnel. The spring-tides, rushing up the Channel, enter an opening of some eight miles' width, the steadily contracting banks of which squeeze together and pile upon one another the waves, in their upward course, till a crest or bore, of from twelve to sixteen feet in height, according to the direction or force of the wind, is formed, which surges up the river in a single step. At the equinoctial spring-tides this *flôt* is regarded with considerable apprehension; and a vessel unprovided with a Quillebœuf pilot, especially if stranded (which would, under these circumstances, be pretty surely the case), would probably 133 founder with the shock. The slight differences in the tidal force, and in the action of the wind, that distinguish each successive spring-tide, cause the sand, which is disturbed and whirled about by the force of the *flôt,* to be formed into fresh and unexpected shoals, month after month. Such, at least, was invariably the case at the time in question. During the *mortes-eaux,* or the two or three days of neap-tide, navigation is almost suspended; the sluggish rise of the waters being insufficient to bear the sailing traders over the sand-banks, and the services of a steam-tug being indispensable, while the channels are often too narrow for the advantageous application of steam towing-power to more than a single vessel at a time. During these dead neaps, the pilots are usually engaged in sounding the banks and channels formed by the preceding springs. With full spring-tide, and with favourable wind, the crest of the *flôt* is followed within half an hour by a numerous fleet of vessels of all sizes, propelled by steam and by sail, which anchor at the spot where they lose the advantage of the tide, and rarely make an uninterrupted way to Rouen. This famous city, joined to the ocean by so capricious a link, could not compete with Hâvre as a port, even in times when inland water carriage was the great desideratum of commerce. With the revolution in the celerity and the cost of inland traffic, effected by the railway system, the most seaward points of departure resumed the importance which they had possessed during the period of the timid navigation of the Italian rulers of the world, and serious competition between Rouen and Hâvre became out of the question. This clear appreciation, however, was not evident to the Hâvre merchants in the year 1842.

It was thus a very natural circumstance, that those residents of Hâvre who were most interested in the maintenance of the commercial importance of their port and docks, should avail themselves of the advice of an Engineer, fresh from the construction of the only railways then actually in operation, as to a steam communication with Paris. And so serious a form did the negotiations assume, that the Engineer in question was authorised to rely on the formal support of the town of Hâvre, if a certain amount of aid could be depended on from England.

It was generally understood in the profession at this time, that Mr. Robert Stephenson did not view without some displeasure the rapid rise of Mr. Locke,[77] and especially the establishment of that Engineer in France, where he had introduced, on the Paris and Rouen Railway, the method of construction by preliminary bargain with a single large contractor. It had, in consequence, been anticipated at the Hâvre meeting, that Mr. Stephenson would readily act either as Engineer-in-Chief, or as Consulting Engineer, to the proposed

[77]George Stephenson had begun to dislike Locke in 1829 when Locke, at the request of the Liverpool and Manchester Railway board, reported on the Edge Hill tunnel and was obliged to speak critically of the work that had been done. The breach was widened in 1834-35 when Locke was made engineer of half the Grand Junction line, independently of Stephenson, and then sole engineer. Robert Stephenson supported his father at first, but was in the end cordially reconciled to Locke. In this case, as in most others of the kind, George Stephenson showed himself implacable.

line. The younger Engineer, therefore, on returning to London, went immediately to Great George-street, where he was received by Mr. Stephenson with the frank cordiality with which that gentleman, in those days, always met the younger members of his profession. The whole scheme was produced and discussed, the names of the termini being only reserved until it was seen how Mr. Stephenson would treat the proposal. "It looks very fair," said he; "the traffic must be ample to warrant the construction of such a line as you describe. I am very fully occupied, and could not devote much time to works in France. But I should rather like to open a connection there; and if, on further examination, your statements are verified, and if you can undertake the responsibility of discharging the duties of Engineer-in-Chief, I shall be very happy to give you my support and advice as Consulting Engineer. You can easily understand how far I shall be able to go into the details of the affair." "Nothing could be more gratifying," said the other; "it is more than I could have asked. The line is from Hâvre to Rouen."

Mr. Stephenson's countenance fell. "I am very sorry to hear it," said he; "sorry, I confess, on my own account, and still more sorry, 135 after what I have just said, to disappoint you. It is the only spot on the Continent in which I cannot do what I proposed. You are, very likely, aware that there is an unfriendly feeling between ourselves and Mr. Locke. We think he has used us ill: whether we are right or wrong is not to the point. But, under that impression, I feel it due to myself to avoid giving him any possible ground of complaint against me. I know that this is not like running an opposing line. Mr. Locke has no right to monopolise French practice. Still, the line to Hâvre is so necessary a prolongation of the line to Rouen that he might say I was interfering with his district, if I went there, while all the Continent was open to me. I must decline. I am the more sorry because I cannot recommend you to go to any one, who can do what I could do for you in the matter, but I see no alternative.''

[78]Can this young engineer have been Conder himself? He gives a circumstantial account of the interview with Robert Stephenson, and the phrase 'so great a disappointment' might seem to suggest it.

It was impossible, even under so great a disappointment,[78] not to honour the feeling which prompted Mr. Stephenson to decline such an introduction to the French practice of which he was confessedly desirous. After a little more conversation, the project was taken to another Engineer, who had before this time made much out of the overflow of Mr. Stephenson's business. This gentleman welcomed the idea with alacrity, instantly proceeded to demand the fullest details, and within a few days gave such unmistakable proof of his readiness to fill the post, not only of Consulting Engineer, but of Engineer-in-Chief, Resident, Sub-Surveyor, and everything else (by himself or his staff), that the promoter, determined not to act as a mere cat's-paw, declined further interference in the matter, and the line, in the following year, fell ripely and naturally into the hands of Mr. Locke.

The unpleasant feeling between the Engineer of the London and Birmingham, and the Engineer of the Grand Junction Railway (lines quite as much functions of one another as were the Paris and Rouen, and Rouen and Hâvre), was subsequently removed. Mr. Locke came down to Chester, on the occasion of the inquest into the Dee Bridge accident, to support Mr. Stephenson. A little incident at a somewhat later date showed the pleasant relations of the two men, both with each other, and with their frequent opponent, Brunel. The three were travelling together in a railway carriage; Stephenson wrapped in a dark plaid, on the exact disposition of the folds of which, he somewhat prided himself. He saw Brunel regarding him with curious eye. ''You are looking at my plaid,'' said he. ''I'll bet you ten pounds that you cannot put it on properly at the first time.'' ''Very well,'' said the other; ''I have no objection to bet ten pounds. But I won't take your money. I bet ten pounds against the plaid. If I put it on right when we get out on the first platform, it is mine. If I miss, I pay you ten pounds.'' ''Done,'' said Stephenson, and resumed conversation with Locke. But Brunel sat in a brown study, and said not a word till they arrived at the next station. "Now then, Stephenson, give me the plaid to try," said he, as he stepped on the platform. Robert Stephenson slowly unwound the garment. Brunel promptly wound it around his own shoulders, with as much composure as if he had pulled on a great coat. "It is a first attempt," said he, "but I think the plaid is mine." For many a day did he rejoice in its comfort.

"But had you never tried before?" said a friend. "No," said Brunel; "but when Stephenson challenged me, I was not going to give up; so I began immediately to study the folds, and to make out how he had put it on. I got the thing pretty clear in my head before we got to the station, and when I saw him get out of it I knew that I was right, so I put it on at once." A playful instance of the concentrated meditation which Mr. Brunel gave to any subject on which he fixed his thoughts.

The Dee Bridge accident, before referred to, was almost the sole blot on the safe and brilliant course of Mr. Stephenson's career. The subject excited great interest at the time, and a scene in which the skill of a practised advocate, and the whole tact and authority of Stephenson and Locke united, were bent to disprove actual fact, was one which hardly needed the additional novelty of the appearance in court of one of the bravest and coolest men in the world, trembling like an aspen, to fix it indelibly on the recollection.[79]

The bridge constructed to carry the Chester and Holyhead Railway over the river Dee, close to the city of Chester, consisted of twelve cast-iron girders, of a span of 98 feet each in the clear, between the piers. Each girder consisted of three pieces, bolted together at the joints, making a length of 109 feet, of the depth of 3ft. 9in. The width of the top flange was $7\frac{1}{2}$ inches, that of the bottom flange 2ft. $4\frac{1}{2}$ inches, the thickness being from $1\frac{1}{2}$ to $2\frac{1}{2}$ inches and the total sectional area 160 square inches. On the 24th of May, 1847, five inches of broken stone ballast were laid on the bridge, which had been for some time used for public traffic. On the passage of the next train, one of the girders broke under the weight, precipitating the train into the bed of the Dee, killing four persons, and so injuring a fifth, that he died on the second day after the fall. An inquest was held which lasted several days, and which assumed the form of a scientific inquiry into the cause of the calamity. The most clear and indubitable result of this investigation was, to demonstrate the entire unfitness of the ancient machinery of the coroner's court, presided over by a country doctor or lawyer, in which solicitors appeal to the convictions of an incompetent jury, unchecked by the presence of a bar, to deal with questions involving such important issues.

[79]For a modern treatment of this accident and of the engineering questions it raised see P. S. A. Berridge, *The girder bridge,* 1969, 9-30. Robert Stephenson's compound beams appear to have been mainly responsible for it.

The scene, in fact, at one time degenerated almost into a personal contest between Sir E. Walker,[80] the foreman of the jury, (who seemed resolved on the return of a verdict of manslaughter, or, if possible, of murder, against the Engineer, or some other unfortunate servant or representative of the Company,) and Mr. Stephenson himself, who betrayed, by his appearance and manner, the most intense vexation and inquietude. It was evident that the sympathy of the man with the loss and pain caused by this occurrence was the emotion most present to his mind, but the disquiet, as is so often the case, betrayed itself by an unfortunate disturbance of the usual manner of the great Engineer. Pale and haggard, he looked more like a culprit than like a man of science, assisting in a painful investigation. His manner was abrupt and dictatorial, betraying extreme irritation at the remarks of the jurors; and on more than one occasion he attempted, on the score of professional knowledge, to put down with some contempt the questions and remarks of Sir E. Walker, who, nettled in his turn, affected to treat Mr. Stephenson as a culprit on his defence.

The case for the Company was conducted by their solicitor, a man well known in the City of London, in which he held a high legal office, being the adviser of successive Lord Mayors, and an authority of great weight in the Guildhall.[81] A compact man, somewhat below the middle size, with close, grizzled locks, and a settled courtly smile, who received the reply of a witness with equal urbanity, whether it went to establish or to demolish his case. His dress was of extreme neatness, though even at that date tending to be old-fashioned: a

full suit of black; dress coat buttoned tightly across a broad chest; no shirt-collar, but a white muslin handkerchief bound tightly, and in innumerable folds, around the neck. With a self-possession which nothing seemed able to disturb, an eye that pierced the recesses of the jury-box, the tiers behind the bench, the thoughts of each witness successively under interrogation, the lawyer presented the aspect of a man at peace with his own conscience, and in loving charity towards all the world, especially the jurymen. He brought forward the bridge-master of Chester, the superintendent of the iron works where the girders were manufactured, Mr. Kennedy, the locomotive manufacturer, Mr. Locke, Mr. Vignoles, and Mr. Gooch, to support the report and evidence of Mr. Stephenson. The line which he selected for defence may be best described by the account of a conversation held with him, on his return to London after the conclusion of the inquiry. ''It was most interesting, Mr. Solicitor, to watch the manner in which you conducted the case. I should like to take the liberty of putting one question to you, now it is over.'' ''I shall hear it with great pleasure.'' ''It seemed to me that there was one point rather unaccountably overlooked. Were you disappointed of a witness, or is it my stupidity?'' ''I do not see to what you refer!'' ''Why, up to a certain point, your defence was perfect. You proved, by scientific evidence, first, that the bridge ought not to have broken; and, secondly, that, at least from any assignable cause, it could not have broken. You only wanted the witness to prove that it did not break, and the case was your own.'' Mr. Solicitor did not quite like it. ''Jesting apart,'' said the other, ''it was impossible not to admire the mode in which you carried out your line of defence; but I should like to know why you selected that line?'' ''What other was open to us?'' was the reply. ''It would have struck me in this way,'' said his interlocutor. ''If you had come forward and said, 'Here has been a great calamity; the bridge which has broken was constructed according to the best scientific information of the day. You have in evidence that a hundred bridges are constructed, or in course of construction, on the same principle; and you have heard some of the most eminent Engineers of the day state that they approved the principle. The Government Inspector-General of Railways has told you that he sanctioned the opening of the line, and considered the bridge to be safe, from his experience of similar bridges on a smaller scale. The failure of the girder of this great span has been the first intimation, to scientific men, that the formula required revision under altered circumstances. No human wisdom could have foreseen such a result. We are in the presence of a most unhappy, but a most instructive, experiment, which will be a safeguard for the future; and the Company will make every compensation in their power to 139 the families of the sufferers. It must be regarded in the light of a casualty in war.' Now, if you had taken this line, I do not see how you could have been defeated; and I should like to know why you rather preferred to fight against the facts." "Because," said the Solicitor, "you should never admit anything, especially with a jury. Between men like you and me, yours would have been, no doubt, the best course, but with a jury never admit anything. You do not know how it may be turned upon you."

Mr. Solicitor's defence, however, was unfortunate in its results. The verdict did not find any one guilty of manslaughter, but it stated the unanimous opinion of the jury, that the girder did not break down from any lateral blow, or from any defect in the masonry, as urged by the defence, "but from its being made of a strength insufficient to bear the pressure of quick trains passing over it." And the jury called on the Government to institute an inquiry into the strength of all similar bridges.

The mechanical law which was illustrated by the serious catastrophe was not at the time of the inquest, nor perhaps at any subsequent time, succinctly and clearly stated. It was no other than that which defines the respective provinces of Statics and Dynamics.

In attempting the construction of iron beams adapted to support weights, rarely dealt with within the limits of modern experience, when moved at great speed, over openings of unprecedented span, the calculations of the Civil Engineer were necessarily based upon experiment. The *formulæ* on which reliance was placed had been deduced from a series of experiments on beams of different dimensions and proportions, which had been supported on fixed bearings, and gradually weighted until broken. It was clear that guidance might be obtained, by calculation based on the *formulæ* thus arrived at, as to the construction of girders calculated to bear a given pressure, and to span a given interval. But, with the rapid motion of the locomotive, it became necessary to provide, to some extent, against the more formidable force of IMPACT. A deflection which might be unimportant when caused by a gradually imposed pressure, assumed quite another phase when caused by sudden pressure, accompanied by great vibration. Nor was vibration the only force by which the new element of motion tended to strain the structure of the beam. In the movement of the locomotive, even over the best adjusted line of rails, there is much greater force exerted on the surface of the rail, when the cranks are in a certain position, than during the remainder of the revolution of the driving wheels. When to this recurrent variation in stress, there is added the consideration of the inequalities that tend to give something of a jumping motion to a train, including the play of the springs, it is clear that a certain amount of hammering is caused by the passage of the great weight which is whirled over a railway bridge. This hammering is distinctly sensible to the observer who stands on such a structure at such a moment. To provide, therefore, for safety, it is necessary to consider, not only the capability of the structure to bear the maximum weight of a train, but its capacity to resist, at the same moment, the utmost dynamical force which the moving parts exert under any possible circumstances. It became evident, from the failure of the Dee bridge, that this had not been done in the case of the girder in question.

The breaking weight of the pair of girders, of which one failed in the present instance, was 120 tons, according to one *formula*, and 150 tons according to another. The weight of the platform, exclusive of that of the girders themselves, was about 20 tons. To this 25 tons of ballast had been added immediately before the fracture. Half of this weight according to the *formula*, was to be considered as concentrated on the centre of the girders. To this was suddenly added 32 tons of dead weight of locomotive and tender, a proportion of the weight of the rest of the train, the vibration caused by the passage, and the successive blows of the revolving wheels. The statical resistance of 120 calculated tons of pressure was thus shown to be unable to withstand the sudden shock of the train, when the girder was already weighted with platform and ballast. The practical rule, up to that time, had been to reckon a third or a fourth of the breaking weight as the maximum weight to which the girder should be exposed. In this case the actual stress was as $52\frac{1}{2}$ to 120, according to the lowest *formula*, and the result was not one that was in itself calculated to surprise a student of Dynamics.

But in addition to insufficiency of section, there was a grave mechanical error in the construction of the girder in question, to which, although designed by an assistant who was famous for his powers of calculation, it is unaccountable that the practical eye of Mr. Stephenson should have been for a moment blind. The cast girders were complicated with wrought iron tie-bars, attached to castings bolted to the top flange, which descended obliquely for a third of the length of the girder on each side, and were secured close to the bottom flange for the central third of the beam. Now, if these tie-bars had no tension put upon them, they were not only useless, but by their weight, and in consequence of the weakening of the girder by the holes bored for their attachment, were injurious to the strength of the structure. On the other hand, if tension were applied, either by keying up, or by deflexion of the girder, the direct effect would be to break the beam which the bars were supposed to strengthen; for the resistance of a double T girder to fracture from weight, is calculated on the united power of the top tables to resist compression, 141 and of the bottom tables to withstand tension, or tearing apart. Now, every pound of tension put upon the tie-bars represented a proportionate crushing power applied to the top tables, and thus aided, not the resisting power of the girder, but the destructive power of the load.

The veteran James Walker,[82] together with Captain Symonds, R.E.,[83] who were employed by Government to examine the case, referred in a dubious manner to the probable want of joint action between tension-bars and the girders. They failed to see the possibility of a contrary action. Mr. Robertson, the Engineer of the Chester and Shrewsbury Railway,[84] had the courage to express the opinion that the tension rods weakened the girder, and to oppose the theory of Mr. Stephenson and his supporters, that the fracture had resulted from a blow given to the girder by the tender of the engine, assumed to have been previously thrown off the rails by some unknown cause. Major-General Sir Charles William Pasley, R.E.,[85] the Government Inspector-General of Railways, gave original and curious reasons for disapproval of the tension bars.

General Pasley was a man of whom any service might have been justly proud. The Royal Engineers owe much to his laborious study of his profession. Perhaps no officer or man of science ever put the result of experimental labour in language so clear and precise. It was impossible to misunderstand General Pasley's instructions. In speaking, for instance, of the planks to be prepared to form a gangway for wheeling over, he was not content with recommending that the ends should be bound with hoop-iron (in itself a useful, practical suggestion); he specified the length, width, and gauge of the iron, the number and position of the holes to be punched, and the size of the nails with which the guard was to be fixed to the plank. Probably no such *vade mecum* has ever been constructed as the specifications of General Pasley.

To patient, constant, conscientious toil at the practical details of his profession, Sir Charles Pasley added a courage that was heroic and romantic. This courage was not exclusively military. It was not shown under fire alone, but he has been known to look at death, apparently imminent, under circumstances that might dismay a brave man, with an unconcern that scorned even to smile. It is recorded of him that once, while labouring himself in the trenches, a French assailant pierced him with a bayonet. Pasley, severely wounded, turned on his enemy, and killed him with his *spade.* During the experiments which he was conducting on the strength of cements, a large stone of more than a ton in weight had been cemented to another and the two had been suspended in the air, by a lewis inserted in the latter. Pasley went to inspect the joint with his usual scrupulous and minute care. He looked above, around, from below, walked

[82]James Walker (1781-1862), President of the Institution of Civil Engineers in 1834-45.

[83]Captain J. L. A. Simmons, RE, Inspector of Railways 1847-50; afterwards Field-Marshal Sir John Simmons.

[84]Henry Robertson (1816-88), responsible as projector and engineer for much of the railway system in north-east Wales.

[85]Sir Charles Pasley.

round and under the suspended mass, peering and scrutinising in every direction. Just as he stepped from beneath, the cement gave way without warning, and down fell the heavy block. A second or two earlier, and it would have crushed Pasley flat. Without the slightest change of colour, of tone, or of step, he turned to his assistant: "Do you know," said he, "I call that an extremely unpleasant occurrence. I was just beginning to feel such confidence in the cement; it seemed to behave so admirably: but you see I had no great ground for satisfaction."

Short and compact of stature, with a keen eye, a resolute nose, a complexion bronzed by the sun and storms of many a campaign, General Pasley possessed all the courage of Ney, with much more of the aspect of a Captain. His utter disregard, or apparent non-perception, of danger while prosecuting any scientific investigation, recalls the biography of Maurice of Nassau, who, like the English officer, so well knew the importance of the spade as a military weapon. But an instance of his perfect *sang-froid,* when exposed to the danger of another element, to which a soldier may be supposed less accustomed than to perils of fire, or of earth, may perhaps give a more vivid impression of his imperturbable courage and self-command.

Pasley had descended with a junior Engineer officer in a diving-bell, at a time when he was giving attention to submarine experiment. By some misfortune, the apparatus became deranged, the chain was entangled with the air-tubes and signal-line, the supply of air became strangled, and the water slowly and steadily rose in the bell. It was impossible to creep beneath its edge in order to dive out and float to the surface, for the bell rested on the bottom of the sea, and that was an inclined plane, so that the close dome was tilted considerably to one side. Death appeared inevitable; a question of moments or of inches, as the water slowly crept up the limbs of the two officers, who were thus utterly unable to make any effort for escape. "I confess," said the junior, "that I felt a very considerable alarm. But Pasley never changed a muscle; he continued talking in the quietest manner on indifferent subjects, and watching the rise of the water, as if it were an experiment he was superintending in the ordinary routine of duty. The water had risen to our breasts, when the people above, receiving no signals, became alarmed, and drew up the bell. I cannot describe my feelings, but Pasley took it all as the merest matter of course."

Yet this brave veteran, cool in the storm, in civil danger, in the face of drowning, trembled like an aspen leaf before the great dread of the official in England—Responsibility. This is no mere figure of speech. In October, 1846, General Pasley had inspected the Chester and Holyhead Railway, and had reported the Dee bridge as safe. It had proved to be otherwise. In giving his evidence on the inquest, stating with the most precise detail, and the most unswerving accuracy, what he had reported, and the reasons of his opinion, the agitation of the old soldier was so great, that he was for a time almost inaudible, and it was only the slow, deliberate, manner, in which he brought out every syllable of his evidence, that enabled the bystanders to comprehend what he said. The rough military reason, that as a similar bridge over the Tees, at Stockton, with a span of 83 feet 4 inches (which Mr. Stephenson corrected to 87 feet), and a depth of girder of 3 feet, had proved trustworthy, a bridge of 98 feet span, with a depth of girder of 3 feet 9 inches, might be thought safe, might certainly have served to shield the Inspector-General from any charge of carelessness, in a matter which had just been illustrated by so novel and so fatal an experiment.

CHAPTER XX

A canal railway

"I am sure of one thing," said Robert Stephenson to a party of the Directors of the London and Birmingham Railway, who were accompanying their Engineer in his twelfth walk from end to end of the line. "I am quite sure of one thing; we have taken the right district." That opinion had the sanction of an earlier authority. The Grand Junction Canal had been laid out through the same line of country, with the exception of a few miles of deviation caused by the opposition of the Earl of Essex to the piercing of Cassiobury Park by a railway. But the line in question was not the only one in which the surveys of canal Engineers had indicated the best lines of internal communication. Throughout England the canal was the pioneer of the railway.

The essential difference between the works suited for the successive ponds, or hydraulic steps, of a canal, and the regular gradients of a railway, as well as the greater licence occasionally taken by the constructors of the former class of works for sudden bends, prevented the promise of much economy from a scheme for converting a canal into a railway. The proprietors of the means of internal navigation, moreover, while naturally doing their best to impede their new rivals, were not prepared for such a brilliant success on the part of the latter as to induce them to give up the struggle, and to offer their land and their powers as contributions towards a common capital, to be divided between themselves and the incoming subscribers. Conversions, however, did take place, and it may be interesting to refer to a case in which a canal, which was subsequently thus replaced, was, in the first instance, bordered by a railway, without destroying the actual water communication.[86]

To form and consolidate the towing-path over two long, straight reaches, without a single lock, to cover the surface with Thames ballast readily brought to the spot in barges, and to lay a strong sleeper railway on the same, was a labour remarkable only for the extreme facility afforded by the convenient water carriage. Amicable arrangements with a great landowner allowed a sharp bend to be

[86]The canal was the Thames and Medway, opened in 1824 (engineer William Tierney Clark). It included what was then the second longest tunnel in Britain (3909 yards), from Higham to Frindsbury. In 1844 the company turned itself into the Gravesend and Rochester Railway and laid a line along the towing-path of the tunnel and a timber viaduct cantilevered out over the water. Fox was the contractor, Conder his resident agent. This is the operation described here. The South Eastern Railway bought up the company in 1845, closed the canal, and laid a railway on its bed, opened in 1847. C. Hadfield, *Canals of Southern England*, 1955, 99-101, 275-276.

145

cut off by a diversion, memorable for only two reasons. The first was the soapy character of the soil, which refused to stand in excavation at a less slope than three to one, though the depth was only a few feet. The other was the discovery of Roman pottery in a very perfect condition, cinerary urns, of a brown colour, some of large size, and containing ashes, and small saucers of the size, and almost of the shape, of those used for garden pots, of a fresh glaze and brilliant red, some of which bore, sharply stamped, the word GRACISCAE.

Two or three iron swing bridges over the canal were replaced by wooden substitutes, giving the additional height requisite for crossing the rails on a level. A brick bridge of one arch, through the only instance of deep cutting, was replaced by a handsome iron skew bridge, the erection of which, or rather the preliminary process of demolition, afforded a triumph of ingenuity, in the reduction of scaffolding to a minimum.

The work had been entrusted to a foreman, who was so typical an English workman, that his memory deserves to be rescued from oblivion. A tall, strong, active man, with a face such as that which Shakespeare attributes to Bardolph, fertile in resource, untiring in activity, and with the courage of a lion. A carpenter by trade, and engaged as a foreman and sub-contractor on the works of the Canal Railway, nothing came amiss to him. He had a natural facility for the organisation of workmen, and for getting the most work done, in the shortest time, by the smallest number of hands. With nothing but a punching-machine and a portable forge, and apparatus of his own construction, he would convert a quantity of iron bars into a light and substantial roof; working in the open air until protected from the weather by his own growing edifice. In dealing with weight he showed true mechanical genius. ''People only care,'' he used to say, "*not* to see what is under them. Give them a platform that they cannot look down through, and they are content. They never look at its legs. I like to see what I stand on. When one is certain on this point, one can go to work with a will." The consequence of this mode of regarding the practical problems which he had to solve was, that he rarely had any accidents.

The man was ever on the alert. His rapid eye and vigorous elasticity more than once saved his life. On one occasion a load of timber fell over him, knocking him into the water, from which he emerged, to the surprise of all, unhurt, except a few bruises, having made his arms act as a wedge, at the critical moment, between two threatening beams, from under which he shot forth like a projectile. On another occasion a Sub-Engineer was crossing a very rapid but shallow stream, on a temporary plank bridge. He missed his footing, and fell into the stream, alighting on his feet, in water but a little way above his knees, but swaying with the shock and with the force of the current. In less time than it takes to tell, the carpenter had rushed from the bank into the middle of the stream, charging like a wild bull, and had seized the tottering Engineer round the waist. In the same vicinity adders were abundant. The foreman was being driven rapidly in a dog-cart by his master, who espied one of these venomous creatures, fat and yellow, standing erect in a copse within a few yards of the road, and hissing at the vehicle. Almost before the driver could turn his face, his companion had started up from his seat, swung a heavy sledge-hammer round his head, and whirled it with true aim and deadly force at the adder. But the snake, lither in its movements than the projectile, threw itself in a coil round the stale of the sledge, as the iron head buried itself in the ground, and shot off into the underwood.

An occasional, perhaps more than an occasional, over-fondness for the ale-can, was, as might be expected, one of the weaknesses of a very energetic character. It is to be hoped that he was a good Christian. But in some respects he would have been perhaps more at home in a Mohammedan country, than in one where certain prejudices still linger on the subject of monogamy; for his matrimonial relations were perplexed; and if he had occupations in different parts of the country, he thought a *placens uxor* a proper part of the arrangement for each. Sometimes the different claimants to the title would meet on the same spot, and discord would ensue. Flight was generally the best policy on such occasions.

In the case of the bridge which this very useful man had undertaken to demolish, the proceeding was simple. He offered to take it down safely for a certain sum. This agreed, he explained his plan. He jumped a hole through the crown of the arch with a crowbar, and   147 passed a strong iron eye-bolt through the hole, securing it with a good washer and a balk of timber above. He thus had a means of attachment below the arch, perfectly secure. To this he appended a light platform, and set a bricklayer upon it, to take down the arch by one ring of bricks at a time, beginning at the face. When he had thus safely taken down nearly half of the arch, he began at the other face, and reduced the bridge, brick by brick, to a single rib, just sufficient to bear the weight of the workman. Then he withdrew his man, and pushed the narrow rib over. It fell into the canal, and the work was accomplished. Nothing had to be done but fish up a few bricks.

A more serious, and probably an unique, undertaking was the construction of a timber viaduct through the entire length of a tunnel of more than a mile long. The towing-path, which continued through this large and important gallery, was not wide enough to support the railway, the usual ovoid section of the tunnel being such, as to require more space for the chimneys of the locomotive, and for the corners of the horse-boxes, the two most salient points in the section occupied by a running train that could be obtained without trenching on the canal. It was therefore necessary to lower the towing-path to the water level; to drive a line of piles, fourteen inches square, and fifteen feet apart, from end to end of the tunnel; to drive a corresponding line of shorter and smaller piles on the inner edge of the towing-path; to connect the whole by sills and transoms; and to lay half-balk timber, and bridge rails, on the frame-work. The chief difficulty of the work, was the speed at which it was required to be done, and the necessity of carrying it all on by torch-light. The great facility of water communication and carriage was also partially lost, as it was arranged to draw off the water for thirty days, in order to bore for the reception of the main piles—a solid bedding in the chalk being preferred to the ordinary method of driving by a pile-engine.

The first step was to remove the superfluous chalk from the towing-path. When the Engineer who had charge of the construction was preparing to commence this operation, he was applied to by the Secretary of the Company to allow the work to be executed by the Company's servants. "We have a sale for the material," said the Secretary. "We are accustomed to the work. You may as well save yourself the expense, and let me lower the path and make use of the contents." The Engineer said the only objection was in point of time. The timber work would be ready for fixing by a given date, on the arrival of which the water was to be let off; and it was necessary that the towing-path should be completed by that time. The Secretary
promised punctuality, and the agreement was duly concluded and

put in black and white.

But when some third or more of the time had expired, without perhaps more than a tenth part of the work being performed, both parties began to be in a fidget. The constructor, though relieved from penalties as to delay, by the fact that the cause of obstruction was due to the Company, was anxious to preserve his well-earned reputation for punctuality. The Secretary found that he had made a bad bargain, and that work in the dark, under pressure for time, was something very different from what his former leisurely experience had led him to anticipate. Under these circumstances, the Constructor sent for a Sub-Contractor, who was executing the cutting before referred to. "The Secretary has got into a mess," said he. "I think he will be glad for you to help him out of it. The matter is out of my hands, and I will do nothing to resume the responsibility. But I do not want time to be lost. Go to the Secretary, and see if you can make a bargain with him to finish the work. You will be sure to have to do so, sooner or later." The Secretary accepted the offer, made some arrangement—of the details of which only himself and the Sub-Contractor were cognisant—and the work was knocked out of hand.

But when the Sub-Contractor sent in his bill of £300 odd, matters went less smoothly, and so much difficulty was made as to payment, that the claimant went to the Engineer. "I've come to ask you about that towing-path job," said he. "The Secretary says it is your affair, and not his;" and with that, he began to fumble in his pocket. "Stop a minute," said the other. "Listen to me. Do not attempt to show me any paper on the subject. I will hear all you have to say, and give you my best advice; but if you show me anything that I do not ask for, you may put it out of my power to serve you. The whole proceeding has been perfectly regular hitherto, and nothing must be done to put any one in a false position." For the manner in which a man who wants money will run to the nearest available source, without too carefully considering whether that source be legitimate, is a portion of the daily experience of the conductors of large works.

The upshot was, that the Contractor had to sue the Secretary for his bill, and that the case came on for hearing before the Lord Chief Justice. The principal witness was the Engineer in charge of the construction, who was examined and cross-examined at great length. "Do you tell us," said the Judge, "that the Contractor never brought you in a bill for the work?" "Never," was the reply. "I should not have received it if he had done so, as I had relinquished the undertaking to the defendant, at his own request; but I never was shown anything of the kind." On this, his Lordship, who shared   149 the defect possessed by some long-experienced judges of thinking that he could see what did not come out in evidence, charged the jury. "The trial," said the Judge, "though nominally between the plaintiff and the defendant, is really between the Constructor of the Railway and the Canal Company." In this charge, his Lordship made a statement that was not only unsupported, but inaccurate. So the jury thought, for they found a verdict for the full amount claimed by the plaintiff. When the case was over, the witness asked the plaintiff—"You have done as I told you," said he, "and you see that it was right. Now, tell me, when you came to me that day, in such a state because the Secretary would not pay you, had you not made out some claim against us?" "Here it is," said the other. "I had made out a regular bill to you." "If I had ever set eyes on this before," said the other, "you would have lost your suit to-day."

The settlement of this little dispute, however, was preceded by the completion of the tunnel viaduct, the whole of the erection and framing together of which was finished within the prescribed thirty days. The sides for the main piles were bored by a well-sinker's apparatus; a heavy iron auger, of the diameter of the hole, working through a tube of wrought iron, of the same diameter, and suspended from a frame, was jumped down into the solid material by a lever worked by two or three men, and the *debris* was removed by an iron ladle. The smaller holes on the towing-path were jumped by crow-bars, guided by iron tubes. The whole frame-work had been previously prepared in the yard at the mouth of the tunnel, and the only part on which a chisel had to be placed *in situ* was the notching of the piles for the reception of the bevelled sills. The viaduct answered its purpose admirably.

The execution of this viaduct gave a proof of the manner in which depressing and unwholesome physical influences can be thrown off by excitement and exertion. The atmosphere, in this long water tunnel, was heavy and depressing, merely to walk through. To pass through in a boat drawn by a pony, produced a sense of fatigue and discomfort, often sharp head-ache. It was, therefore, always preferable to encounter the additional length of path, and amount of climbing involved in a walk over the surface, rather than to take the shorter cut. But with the atmosphere rendered far denser by the smoke of torches and the exhalations from a large body of workmen, the Engineer in charge of the work, under the pressure necessary for its punctual accomplishment, found himself able to remain for four, six, or eight hours, without sensible distress, in the very spot which he shunned even to pass under ordinary circumstances.

The density of a London fog is well known. With the increased
consumption of coal that is taking place in Paris, the inhabitants

of that stately capital are beginning to learn that fog is not absolutely peculiar to the valley of the Thames. In Sheffield, in Leeds, in other of our large manufacturing towns, fogs of extreme density at times appear, and the passenger through a hilly Yorkshire town has known what it was suddenly to emerge from perfect obscurity into laughing sunshine. On looking back, a dense vapour was seen rapidly rolling southward—a dark cloud, rising like a wall from the ground, and perfectly obscuring all that lay beyond its rapidly receding edge.

No darkness is so intense as that which is subterranean. From the interior of a tunnel in course of construction, all atmospheric light, even such as is diffused in the darkest night, is excluded; and if artificial light be also absent, the pressure of the gloom on the optic nerve is almost appalling. When the mouth of the structure is open to the day, the case is different, as a pencil of rays is then apparent. But a river fog in a canal tunnel has been known to blend the rayless gloom of subterranean darkness, with the palpable thickness which loads the lungs in a London fog. Mechanical obscurity is added to absence of light. So heavy was the veil of fog on one occasion, in the tunnel now referred to, that the bearer of a lantern, containing a wax moon, such as those burned in carriage-lamps, was unable, by its light, to see either hand or foot, without almost touching the member with the lantern. As each step was taken along the tow-path, encumbered with timber and with tools, and skirting the deep and perpendicular edge of the canal, the lantern had to be held within six inches of the ground at each step, before it could be ascertained where the foot could be safely planted. The effect was not altogether unlike that produced by a thick fall of volcanic ashes—it was darkness that could be felt.

A similar, although a more transient, darkness, was produced by the first passage of a locomotive through the same tunnel. On the completion of the viaduct, one of the patent six-wheel, long boiler, locomotives, constructed by Robert Stephenson and Co.,[87] which had been sent by sea to the spot, and put together on the line, was slowly driven over the novel structure. The passage was a new kind of exploration. The accuracy and finish of the timber-work and plate-laying was thus, for the first time, tested, and the section of the tunnel, which varied from length to length, was examined by means of the passage of a wooden frame, which gave a clearance of six inches beyond the most prominent angles of any of the carriages, the top of the horse-boxes being one of the most awkward projections. The slow passage that was dictated by prudence was enveloped in a cloud of steam, which from time to time blinded driver, stoker, and Engineer. At other times the glare from the open fire-box, on the reflecting surface of the vapour, gave a strangely infernal lustre

[87] A design developed by this firm from 1841 onwards, in which all the wheels were placed in front of the firebox. See Warren, *A century of locomotive building*, chap. xxvi.

to the exploring party. It was well that they went slowly. Half-way through the tunnel, the funnel of the locomotive came in contact with a piece of brick-work. It was necessary to unscrew the funnel, to lean it to one side until the obstacle was passed, and to make the necessary corrections in all parts before the return trip. It was to ensure the safety of the passengers through this Stygian pass, where a head too curiously thrust out of a carriage window might have come into fatal collision with the wall of the tunnel, that the expedient was adopted of fixing brass bars outside the windows, a provision intended to be temporary, but which has been adopted to the permanent discomfort of the passengers, on certain Kentish lines of railway.

CHAPTER XXI

A Government inspection

That visit of the Government Inspector, which was a preliminary to the opening of railways in England twenty years ago,[88] was regarded by the Civil Engineers engaged in their construction, as something between business and frolic. Men, educated in the military habits of the old school of Royal Engineers, had little means of obtaining any practical acquaintance with the railway system, except such as were afforded by these visits, made for the purpose of inspection. It would have been unreasonable to expect any great amount of instruction from these opportunities. The time was limited, and a day, or a portion of a day, was all that the engagements of the Inspector could allow him to afford in each instance. The Engineer-in-Chief of the line would make a point of attending in person, with those of his staff who were most familiar with the scene of operations. If any weak points should occur, interesting conversation was wont to assume a particular brilliancy at the moment; and a question of speed of travelling, or of some new modification of locomotive detail, might come on the tapis, just as the party approached any imperfectly finished portion of the line. Shortly after the first locomotives were placed on the London and Birmingham Railway, a scientific civilian, who had given very positive evidence before Parliament, as to the injury to health and other intolerable evils that must arise from the construction of tunnels, paid a visit to the line.[89] The Resident Engineer accompanied him, in a first-class carriage, over the newly-finished portion of the works. As they drew near Chalk Farm, the Engineer attracted the attention of his visitor to the lamp, at the top of the carriage. ''I should like to have your opinion on this,'' he said. ''The matter seems simple, but it requires a deal of thought. You see it is essential to keep the oil from dropping on the passengers. The cup shape effectually prevents this. Then the lamps would not burn. We had to arrange an up-cast and down-cast chimney, in order to ensure the circulation of air in the lamp. Then there was the question of shadow;''—and so he continued, to the great edification of his listener, for five or six minutes. When a satisfactory conclusion as

[88]This appears to imply that such inspection had ceased by the time Conder wrote, but in fact it remained obligatory.

[89]Probably Dr Dionysius Lardner (1793-1859), who had testified against the Box Tunnel in 1835.

153

to the lamp had been arrived at, the learned man looked out of the window. "What place is this?" said he. "Kensal Green." "But," said the other, "how is that? I thought there was one of your great tunnels to pass before we came to Kensal Green." "Oh," replied the Resident, carelessly, "did you not observe? We came through Chalk Farm Tunnel very steadily." The man of science felt himself caught. He made no more reports upon tunnels.

The officer who inspected the Canal Railway is described in another page. Conscientiously minute in his inquiries, he was still rather new to his civil duties. The Engineer-in-Chief was a man who came to the front rank of his profession, rather in virtue of parliamentary tactics, than of any other form of introduction; a white-haired man, of unusual plainness of speech, and singular determination of will.[90] He was wont to do his own work in his own way, and to show little deference to any one else. On one occasion, a meeting of some of his Directors was held at his office. He told them his opinion on the matter in discussion, and, as they seemed disposed to linger and to talk, he looked up from his desk. "Now," said he, "that is all I have got to say—I am very busy—and I wish to God you would go!" On another occasion, he called upon a Contractor (he used strong expletives), "Here, you sir, what do you mean by putting that timber on the work?—it is not Kyanised."[91] "I beg your pardon sir, it is." "I tell you it is not. Can't I believe my eyes?" "If you will be good enough to step here, sir, I can show you the brand of the Kyanising people. There it is—and here, and here!" "Ah! well, I do not care a single rap whether it is pickled or not—it's all rubbish in my opinion; but it is down in the specification, and I am not going to let you do me."

The attention of the Government Inspectors had been specially called to the tunnel, not only on account of the novelty of the wooden viaduct, but by reason of the fact, that it was only lined with masonry for a short distance from either end, and through a few intermediate lengths, where the material was apparently less firm than for the greater part of the bore. It was cut through solid chalk, which for the most part showed a fair and regular section. But in some places, slips had occurred during the original construction, and lofty domes or galleries ran far above the ordinary level of the crown of the arch. To test the stability of these portions, General Pasley adopted the original, and strictly military, expedient, of firing wooden plugs at the roof. A corporal and small party party of sappers and miners, with a mortar and ammunition, awaited the inspecting party at the mouth of the tunnel. A large barge had been prepared, on which a sort of pulpit had been constructed for the inspector, by standing on which he could, with a small mahogany-handled pick, also pro-

[90]John Urpeth Rastrick (1780-1856). See Editor's Introduction, p. vii.

[91]Kyanising was a process of pickling timber in a solution of corrosive sublimate, used for a short time on railway sleepers; so-called from its inventor J. H. Kyan (1774-1850).

154

vided for the occasion, tap the crown of the tunnel as he advanced. The barge was towed by a party of men, furnished with torches. The battery was placed in the fore part of the barge, and Bengal lights were fired from time to time, which brought out into strong shadow each irregularity of the masonry or excavation. On arriving at the first dome-like cavern, the General began to shout, and then in a moment to swear. "Stop, stop! How can I tell what is going on here? Stop, I say!" The Engineer in charge was prepared for the emergency. He had constructed the platform on which the General was stationed so as to slide upwards when necessary. "Now, boys," he shouted, "hoist up the General." Up shot the platform, with the officer still in the height of his wrath. But finding himself thus unexpectedly in motion, and brought up to the lofty top of the rift, while the whole scene was suddenly illumined by a Bengal light, the old soldier laughed so heartily as to shake the barge itself. He was greatly delighted with the joke, or rather with the well-adapted nature of the expedient, and specially referred to it in his report to Government. He satisfied himself by firing three discharges in

different parts of the tunnel, and arrived in high glee at the eastern terminus, no results except a deafening roar having attended his military test.

"How shall I go back, General?" said the Engineer in charge of the works, after the last station had been walked over. The Inspector, who had made his way down partly on foot, and partly on the footboard of the locomotive, stepped quietly into a first-class carriage for the return. "Go?" said he. "How? Go like the devil!" It was a rude test for the tunnel viaduct, but none too rude. A snug *parti carré* at the fish dinner concluded the inspection, to the satisfaction of all parties. The Engineer-in-Chief of the line came out in the fullest force. Anecdote succeeded anecdote, and some of them were of such a nature as to leave, even at this distance of time, a doubt as to how far they were faithful accounts of real occurrence, or how far the good wine might have lent them a brighter colouring.

"I remember," said Mr Rastrick, his face and eyes glowing under his snowy hair, "when we were making the Brighton Railway, that we were in great want of water at Red Hill; so I sank a well there. We had to go to some depth—hundred and fifty feet I think—then we came on a very fine spring. The water was abundant and rose to the surface. It was perfectly clear, and unusually sparkling. I was very much pleased with the result of the well. But by——it turned out something like a well of soda-water. When we put it into the engines, it would not stay in the boiler. It all fizzed out as soon as it was heated. It was impossible to work the engines with it—it was all 'br-r-r'." "Was there any sulphuretted hydrogen in it?" was asked." "No; nothing could be detected in it but oxygen and hydrogen; but I always had a notion that it was not water, that there was an over-dose of one of the constituent gases, and that it decomposed into gas, and not into steam. I dare say the well is to be seen now. I had to bring water from another source at some expense."

So novel and unprecedented was the growth and development of the English railway system, that the inspection which the Government thought it incumbent on them to make, was a very vague and tentative affair. It was a curious duty in which to employ officers, who were accustomed to receive orders of the utmost precision. For the most serious question of the time, the strength of girder bridges, the only available information was to be found in the tables and formulæ prepared by the very men whose work was to be investigated. Slow processes of test, such as are possible in the foundry and fitting-shop, were inapplicable to bridges *in situ*. Careful daily watching, or the rude test of actual experience, could alone give certainty as to the faithful execution of any portion of the works of a long line of communication, in accordance with the drawings. For a Government inspection to have been reliable, it should have been continuous; or, at all events, it should have been occasional and unexpected. To give notice that a line was ready for inspection on a given day, to put everything in the best order for the visit, to exhibit works and workmen in Sunday clothes, was an agreeable homage to propriety. But as to the amount of public safety thus ensured, the case of the Dee Bridge is conclusive. All that General Pasley had there, in point of fact, to say, was, that as somewhat similar works had not failed, he thought the viaduct in question was safe. It was a good military reason. With the time and means at his command, the Royal Engineer was not able to enter further into the subject. He was the man to do so, had it been fairly put into his hands. As it was, he could only obey orders. The futility of those orders was not his fault.[92]

When a Government has decided that it is its duty to apply a test as to the stability of public works, it is clear that the first thing requisite is, to provide its officers with the means of forming a rational opinion. Experiments and calculations should be made, in the first instance, under efficient direction. Imperial safety is not to be compromised by cheese-paring economy. If each Company were made responsible for the solidity of its works, and the adequacy of its working arrangements, in its own pocket and in the reputation of its Engineer-in-Chief, the dread of the possible consequence of failure would probably suffice to keep both up to a fair degree of efficiency. On the other hand, if a real investigation, at once scientific and practical, were conducted under the orders of the Government, the public would know where to look for assurance, and any effort of the Companies or their servants and *employés* to scamp their work, would be efficiently repressed. But to pass Acts of Parliament, containing certain scientific statements unintelligible to most of the legislators; to profess to inspect, in the interest of the public safety; and then to send an Engineer officer, practically unacquainted with the subject, to take a walk, or a drive over the line, was a method which combined several evils. It lessened the sense of responsibility, where alone that sense was available for public security; it took the onus from the Companies, while it assumed no responsibility on the part of the Government; it placed the officers of a scientific corps in a false position, and tended to break down their high sense of duty by ordering them to take part in a sham. It was but a feature of that unprincipled and uninformed meddling, that failure to grasp leading outlines, that want of faith to either the constructors or the users of railways, to either the shareholders or the public, which has characterised our railway legislation. It is hard to point to the name of

[92]This does poor justice to the Inspectors' reports. The Inspectors (all from the Royal Engineers) did not at first understand some of the problems they were confronted with or the devices adopted for dealing with them, especially in the mechanical department of railway work. But their reports and advice were a most salutary check on the premature opening of railways unfit for public use, and their patient observation produced, as time went by, a body of experience that was valuable to the railway companies too. In early days their reports on every new line were published in the *Parliamentary Papers*. A large selection from the later ones is in the records of the Board of Trade (Public Record Office, Kew, MT6, 29).

157

[93] The first Earl of Redesdale (1805-86), Chairman of Committees in the House of Lords, a keen observer and often a pungent critic of railway practices.

a single member of either house, with the honoured exception of that of Lord Redesdale,[93] who has appeared to watch the national interest. That interest is, in the long run, identical with the welfare of the shareholders;—but it is in the long run alone. The present state of the railway share market shows to what misfortune the unchecked competition of rival Companies may drive their proprietors. A capital of eight millions, in one instance alone, has been doubled with the sole result of halving the returns.

Legislation was necessary for the introduction of the new mode of travelling. So much being admitted, the first thing that any statesman would have thought necessary was, to lay down the principles of that legislation. But this was not done. It was thought fit to deal with each case on its merits, that is to say, to open the door as wide as possible to every description of intrigue, of rivalry, and of juggle. The result is to be seen in the present price of what might have been, by this time, a national property of the utmost value. And certainly, whoever else may be to blame, the chief responsibility for profligacy of expenditure rests on the legislature, which not only sanctioned, but encouraged, a frantic and unmeasured rivalry.

CHAPTER XXII

Disputes as to payment for works

When the works of a railway have been completed, the working stock provided, the staff appointed, and the approval of the Government inspectors signified, it might seem that nothing else was requisite before the opening. One other item, however, has to be borne in mind—payment for the work done.

With the larger and wealthier Companies, where the capital was duly provided by calls from the shareholders, and by the issue of debentures, questions of this nature were rare. Payment on account was made, on Mr. Stephenson's works, by monthly certificates, and on those of Mr. Brunel according to fortnightly recommendations. The latter system had the advantage of requiring only half the amount of floating capital, (in money, that is to say, and exclusive of the outlay for plant and stock), on the part of the contractors, as compared with the former; and to that extent was an economical arrangement for the Companies. But as the character of railway enterprise altered, as speculation became dependent, rather on the discounting of future hopes, than on the provision of anticipatory means, Directors came more and more to lean upon their contractors.[94] Certificates were not met, debentures were offered in place of cash, and, in many instances, works were undertaken, (at proportionately augmented prices), which were to be paid for, not in cash, but in the shares, or bonds, or *promises*, of the Company by which they were constructed.

It is natural for the Engineer, to whom application may be made for assistance in the conduct of an enterprise, to render any aid he can to the Directors, even if not exclusively of an Engineering nature. The one sole, great, Engineering difficulty, want of cash, naturally comes before him very frequently. Thus it became not unusual for an Engineer to seek for, or to introduce to, his employers, wealthy contractors, who would be able to execute works on credit. It is clear that this must always be a costly plan for a Company. It is also true that it is one which is unfavourable to the independence and proper position of an Engineer. Still, when the choice lies between the abandonment, or the postponement, of works, and the services

[94]Contractors played an increasingly important part in the financing of railways from about 1845 onwards. The financial crisis of 1866 bankrupted a number of them, and their part in the business became less conspicuous after 1870, though it continued to be important on some railways (see note 113). The subject is treated in H. Pollins, Railway contractors and the finance of railway development in Britain, *Journal of Transport History*, 1957-58, iii, 41-51, 103-110.

of a leviathan contractor who, on his own terms, will do what the Company have not the means of doing, economy and independence are apt to go by the Board. To keep the agreed schedule of prices as low as possible, and to assume, on parchment, and even on the ground, as much power as the weight of personal character can command, is all that is left to an Engineer, who thus places his clients in the hands of a contractor.

A railway, to which we have before referred, was constructed by a well-known Engineering firm, at a schedule of prices, payments being made in the bonds, or preference shares, of the Company.[95] The work was completed, the stations were built, the locomotive and carriage stock was provided and put on the line. The final measurements were all agreed, the final certificate was given by the Engineer-in-Chief, the report of the Board of Trade was favourable, the Secretary of the Company issued his notification of the running of the trains, and the Directors and contractors met, at the offices of the Company, to settle the question of final payment.

These matters are often regarded from very different points of view. That Engineer alone is entirely independent, who carries out the stipulations of the contract, as to measurement, and certifying for work done, without reference to the pecuniary arrangements of the Company. Such an independence may be rare; but economy and excellence of work, to say nothing of private self-respect, are endangered, when it is infringed. In the present case the Engineer, who had been called in simply to preside over the execution of the works, and had no permanent connection with the Company, did his plain and exact duty, neither more nor less. But the Secretary was less able to deal with the requirements of the contractors, than the Engineer had shown himself ready to certify their propriety.

Companies, somehow or other, are always exceeding their powers. It sometimes happens that the contractor, with more or less justice, may expect that the whole available capital, or power to raise capital, should be applied to the satisfaction of his claims, and that Chairman or Secretary have certain other objects for which they seek to establish a lien on the paper capital. In the present instance the certificate of the Engineer was to a larger amount than had been anticipated, not by the contractors, but by the Secretary; and difficulties arose as to the mode of payment, or even as to the immediate recognition of the certificate. Patience and good temper might have set all straight, but one of the contractors, a man of violent and imperious spirit—who, though able to win, when he chose to make himself agreeable, generally chose to drive, and thus succeeded in becoming almost universally detested—rose from the table with harsh threats to prevent the opening of the line.

[95]This must be the Gravesend and Rochester Railway, from the references to "the tunnel", the "cathedral city" and "the canal" (pp 161–163). The "well-known engineering firm" was Fox's.

160

Discussion for the day was at an end, and the Engineer who had taken the entire charge of the construction of the line instantly took post-horses, and made for the spot. He selected the terminus farthest from London, as that which promptitude would enable him to reach without any companions despatched by the Secretary, or the Solicitor of the Directors. He arrived in the evening, cautioned all the people to avoid violence, in the event of an attempt of the Company to seize the works, and took certain steps of a simple, but perfectly effective character. He then ordered dinner, and quietly awaited the event.

Some hour or two afterwards, post-horses arrived at a gallop, and the Secretary and Solicitor of the Company, one or two of the Directors, and the new Superintendent of traffic, arrived in hot haste and much dignity. They had taken the public conveyance to the terminus of the line nearest London, had proceeded to seize the station, to put their own policemen and officers on the line and to announce the opening and regular service of the trains, for the following morning. They then posted to the further terminus, where the principal station was erected, and in their turn ordered a triumphant dinner.

Their hilarity was disturbed by the entrance of the Superintendent, with scared looks. He had been to the station in order to see to the decoration of the Directors' train, for the following morning, with a proper supply of laurels and of flags. But none of the locomotive engineers[96] (men in the service of the contractors) were to be found. By some unfortunate coincidence they had all gone for a holiday, and had omitted to leave word where they might be found. The telegraph in those days was not available, and it seemed doubtful whether, by any exertion, new enginemen could be found in time for the early opening. An express, however, was sent off to *borrow* an engineer or two, and the toasts were resumed, with the comforting assurance, that the delay of the Directors' train for a couple of hours was the worst *contretemps* to be dreaded.

Later came worse news, however. The Superintendent had returned to his labours at the station; and when just about to depart in comfort, had his attention called to the circumstance, that all the engines were drawn up in single file in the tunnel. He desired a fire to be lighted under one of them, and then made the further discovery that the ends of the steam cylinders were absent. One of these essential portions had been removed from each engine, and not only was steam locomotion thus rendered impossible, but the line was so blocked, that the carriages could not even pass if propelled by horses.

The intelligence had a disconcerting, not to say a sobering, effect on the triumphant seizers of the line. The evening had been, it is possible, too convivial for them to appreciate with exactitude the

[96]Meaning "engine-drivers": a usage accepted in America but rare, by this time, in England. The *Oxford English Dictionary* has no example of it from England at all.

ridiculous position which they occupied; and there was little to be done, except go to bed.

By early daylight on the following morning, the hostile Engineer, who occupied rooms in the same hotel with the Directors, having learned that the police were engaged in an active search for the missing cylinder-ends, thought that they might possibly look under his bed. He had a post-chaise quietly made ready in the yard; and then sending half a dozen strong men privately to his room, deposited the iron-plates snugly in the bottom of the vehicle, and sent off a resolute foreman, in charge to convey them to London. But the Solicitor had taken the precaution to establish a vedette outside the town, and the post-chaise thus stationed, seeing another pass at speed, gave chase. The pursuer came up with the fugitive in the river, where he was about to board a steamer, handcuffed him, on charge of theft, and brought him and the property back in triumph to the Cathedral City, where the Directors were still in perplexity.

By this time, news of the squabble had spread far and wide. The hotel was surrounded by a mob; the railway station was surrounded by a much larger one. The city Magistrates met in session; and when a couple of scarlet jackets came up at a gallop, and the stern features, (which resembled those of Lord Brougham almost as faithfully as a portrait), of the man who had struck the lucifer that had kindled all this flame, looked out, it was rumoured that the troops had been sent for to prevent, or to quell, a riot.

The unfortunate foreman who had been arrested with the cylinder-ends in his possession, was at once brought before the Magistrates on a charge of felony. The contractors, on the other hand, appeared to claim the portions of the locomotives as their property, and to justify their servant as acting under their orders. The lawyers, now quite in their element, exaggerated the strife. The Magistrates looked on with a droll aspect of unusual importance, as the guardians of the public peace, and the dispensers of justice between two parties, of much higher importance than the offenders against the quiet of the streets, or the safety of the neighbouring orchards, as to whom their jurisdiction was chiefly invoked. ''Mr. So-and-so,'' said the thorough-paced man with the stern visage, ''couldn't you contrive to slip out and regain possession of the station, while I keep the Magistrates in play? They are quite at their wits-ends.''

Not only the wits, but the session of the worshipful Bench was, however, overturned by the next hot-footed messenger. Seventy navvies, armed with crowbars, picks, and shovels, had just issued from the tunnel; and, taking the Company's men, who were in possession of the station, unexpectedly in the rear, had secured the traditional nine points of the law to the contractors. The Magistrates broke up the Court, and proceeded to the spot, sending fresh and urgent applications for support to the neighbouring barracks. Danger, however, existed only in the imagination of their Worships. On arriving at the ground, and seeing the actual position of the case, the shrewd common-sense of the Solicitor of the Company decided that there was but one feasible issue from the false position into which the belligerent propensities of the Secretary had carried the Board. A word or two was exchanged between the Solicitors, the result of which was that the Magistrates were requested, by both parties, to assume the possession, and the responsibility of the safeguard, of the line and stations until all disputes were arranged. The chief elements of strife were thus diminished by mutual concession; the basis of a fair arrangement was agreed on; and all parties finished the day by a dinner, which was not only well spread, but well earned. Peace was definitively concluded; and an hour or two of calm discussion in London, on the following day, resulted in the settlement of the account, and the handing over of the line to the Company. The Secretary and Superintendent, on opening the line to traffic, nearly celebrated the event by a human sacrifice, running the first passenger train with such impetus into, and through, the terminus, that the leading wheels of the engine actually left the rails, and hung over the deep basin of the canal, into which a pound or two more of momentum in the train would have precipitated the locomotive, and probably dragged the passenger carriages. The inexperienced officers were consoled by the reflection that a somewhat similar accident had attended the opening of the Euston-square Station, where not only confusions and alarm, but the loss of several teeth, convinced the Engineers that they had not altogether appreciated the force of gravitation, on the sharp gradients falling from the Regent's Canal.

163

CHAPTER XXIII

Railway work in Ireland

It has not been unusual to speak of Irish jobbery. The repeal of the Union is referred to as a successful treaty between the English minister of the day, and those M.P.'s who thanked God that they had a country to sell. People who live in glass-houses have been long warned of the danger of throwing stones. How is it that English railways have cost exactly three times as much per mile as those of Ireland? It certainly is not because they contain three times the quantity of work. The earthwork and masonry executed between London and Bristol are very light as compared with those executed between Dublin and Cork. The cuttings through solid rock on this line find few to parallel them in England. The Boyne Bridge will compare respectably with most of our railway viaducts, if we except the giant strides taken at the Menai, at Chepstow, and at Saltash. And certainly, howsoever much it may be thought that the estimates of the Engineer should be reduced, if the price of agricultural labour in the district he has to traverse is low, no experience in high-priced England can lead anyone to anticipate the difficulty that seems inseparable from the employment of Irish labour. The English employer has to provide for one indispensable requisite alone—if the pay is regular, all, for the most part, goes well. Let the same man go to work in Ireland, and he will do well to provide himself with a fire-proof house, and with a bullet-proof skin.

Not that any one who dwells for a time in the Green Island can fail to be attracted by the charm of the Irish character. Where can we find such ready hospitality? Where such genuine courtesy? Neither at Paris, nor at Brussels, nor at Naples, nor at Turin, nor at Lisbon, (much more not at London,) does the visitant of the city, or the temporary occupant of an hotel, find the tone of constant and unobtrusive good breeding that meets you at Dublin. A man enters Ireland for the first time with his anticipations formed by Lever's novels, or from some slight personal knowledge of "The Mulligan." He descends to breakfast with the idea that he must look to his pockets, and be particular where else he looks, lest he should affront Sir

Lucius O'Trigger. He finds himself among a set of strangers, whose manners make him think Frenchmen theatrical, and Englishmen vulgar.

All this, however, does not prevent a certain sense of insecurity, which rather increases than diminishes with a lengthened residence. The true blot has not yet been hit. We have heard much political eloquence on the subject of Ireland. The worst of political eloquence is that it is true rhetoric—it seeks not truth but persuasion. Residence in Ireland is like residence in Portici, charming while it lasts, but liable to the outbreak, at any moment, of an uncomprehended element of danger.

"Ye need be under no alarm in coming among us," said the manager of a branch of the Bank of England to an English Engineer, on his arrival to commence some considerable works in the South of Ireland. "If ye don't meddle with land, and don't interfere with labour, no one will shoot ye, unless it's quite be mistake." "As we cannot make railways without taking possession of land, and the employment of labour, your's is a particularly reassuring welcome," was the reply. It may be mentioned by the way, that the bank in question had the advantage of keeping its own beggar, a tall lame man with a big stick, who always stood inside the outer door of the establishment, and begged regularly of the customers.

It is rarely the case that a country owes more to an individual than Ireland owes to the memory of Mr. Dargan.[97] He was the true father of Irish railways. No contractor, no dozen contractors, in England ever filled such a position as he did in his own country. No doubt he built his own fortune, although even that proved to be of not much more enduring reality than much of the fair gold that was heaped up in tens and in hundreds of thousands by so many English contractors, only to prove, in the hour of trial, to consist of withered leaves. But he was a man fair in his dealings with all,—just to the Companies that employed him, just to the men whom he employed. His enterprise and energy taught Irishmen to labour at home as they are wont to labour out of Ireland, and formed the mainspring of the Irish railways.

He was singularly modest and quiet, in abode, in dress, in manners; and not only so, but in his estimate of himself. An English Engineer who had undertaken, somewhat too rashly, engagements in Ireland that threatened to prove too heavy for his strength, called on Dargan with reference to some interference which was desired on the part of the Lord-Lieutenant. The Irish contractor saw the point, saw what might be done, but confessed himself ignorant as to whether his Lordship could, or would, do what was requisite. "I came to ask you," said the Englishman, "if you cannot spare the time to go

[97]William Dargan (1799-1867). See Editor's Introduction, p. vii.

165

to Dublin with me, to write a letter to the Lord-Lieutenant, to say that you have talked over this matter with me, and that you strenuously recommend him to do what I ask, as it is of great importance to the prosperity, and even to the peace, of the country." "What!" said Dargan, opening his eyes in unaffected surprise. "Me! me write to the Lord-Lieutenant, me dear Sir! I daur not. I never took such a liberty in me life."

The lovers of Ireland would have done well to present to Mr. Dargan a butt or two of that pure Amontillado sherry, the habitual use of which is the best protection against the undue use of inferior wine. He was killed by Irish sherry. He would stop, in his rapid drives across the country, to change the horse in his gig, and would drink a bottle out of the neck for refreshment. Human nature has its limits, and the well-knit frame, and well-organized brain, of the active man, were unable to withstand this method of keeping up the steam.

It is melancholy to reflect that, even since these lines were penned, an appeal has appeared in the English journals, on behalf of the destitute widow of Mr. Dargan.[98] Under every possible aspect the matter is one for regret. Englishmen in any way connected with the profession of the Civil Engineer, with the business of the railway contractor, or even with the proprietorship of railway stock, cannot see the names, and the appended sums, that made a feeble response to the appeal, without a blush. That any Irishman, who is not by profession a mendicant, should not have felt called on, in the name of patriotism and of self-respect, to do his utmost to prevent such an appeal from being made to the sister country, above all at a moment when Irish politics have been made the engine for bringing public business to a dead lock, is something neither to be explained, nor to be comprehended.

The experience of Irish work which was acquired by the writer of the present pages, was of a nature that may be considered as typical of the difficulties with which the physical reformer has to contend in that beautiful country. The long struggle of the famine was drawing to a close; the public and the private aid of England had enabled her poorer sister to tide over the worst of her sorrows. Men were no longer to be found starved to death on the roadside in the West of Ireland, or at least the instances had become comparatively rare. The attempt to employ pauper labour on a large scale in public or private works had broken down with the most signal non-success. The proverb, "Help yourself, and your friends will help you," had been replaced by the practical lesson, "Show yourself to be perfectly helpless, and then some one will help you, for God's sake." At no time, and under no circumstances, could the prosecution of a large public work, such as a railway, by the

[98]She was granted a civil list pension of £100 in 1870, three years after he died.

stimulus given to the employment of labour and to the circulation of money, have appeared more indisputably in the light of a local benefit and of a national service.

It was not so regarded in the West of Ireland. At least, if there were any who did so regard it, they were few. The Lord-Lieutenant of the country[99] and his accomplished son; the county Magistrate, who thought it not unbecoming the lustre of an old historic name to enter the service of his country as the chairman of a railway company; one or two military men; one or two country gentlemen; —seemed to exhaust the list of those who regarded the opening up of the country by railway communication, in any other light than as a possible occasion for the gratification of individual greediness; and the mode in which this feeling was evinced was after the fashion of the rustic, who killed the goose that laid the golden egg.

No small degree of resolution, of perseverance, and of courage, was required in order to carry on the works. One by one the best English foremen became disheartened, disgusted, or terrified. The inherent tendency of the Celtic race to conspire was shown in a weekly combination against the hand that was feeding them by hundreds. That a man should spend a thousand pounds a week in wages was a proof that he was a fit object for attack. There was something to be got out of him, and if the choice was between getting much by patient industry, or little by idle conspiracy, by menace, or even by robbery, the latter method seemed the most natural and

[99]The fourth Earl of Clarendon (1800-70), Lord-Lieutenant of Ireland 1847–52.

167

congenial to the habits of the people. All the hope of permanently serving the country, the expectation of establishing that feeling of mutual respect and reliance that at times exists between the English owner or employer, and his subordinates, faded before the daily experience of actual life. To complete, creditably to complete, what was actually undertaken, to undertake no more, and to retreat from the country, if not with the honours of war, yet with a skin uninjured by the bullet, the knife, or the club, became the great objects contemplated by those who entered the West of Ireland with the sanguine expectation, that good usage, good pay, and good work, would be appreciated by those to whom they were offered.

It is not so much of the owners and occupiers of the land required for the purposes of the railway that there is room to make any special complaint. It is true that every form of extortion and of petty annoyance was to be met with, but this is not peculiar to Ireland. As a rule, the land arrangements of the railway companies of the United Kingdom have been made, on principles which enlist the instincts of human nature against, instead of on the side of, the purchasing corporation. Instead of interesting landowners to court the approach of a means of communication that must double, or treble, the value of their property, the method usually adopted has been, first, to ignore the local interests altogether, and then to pay two or three times the value of every purchase in consequence. The difference, in this respect, of experience in Middlesex, in Hertfordshire, in Essex, in Norfolk, in Gloucestershire, Worcestershire, Warwickshire, Caermarthenshire, Pembrokeshire, and in the county of Cork, has not been very marked. If Irish tenants have threatened to shoot an invading Engineer, so have Gloucestershire farmers, Essex graziers, and Norfolk parsons. It is true, that in the neighbourhood of Tipperary, the threat seemed to assume a more ugly verisimilitude, but it proved in each case alike a mere *brutum fulmen.* It is not the man who threatens to shoot him whom the Engineer need fear, but the wretch who takes aim from behind a hedge without any preliminary menace.

But the point where the real difference was found to exist was in the state of the labourers themselves. The English navvy has his bad points. Very bad points they are, no doubt, but as a rule they all have a common origin. The fountain of all, or almost all, the troubles of an English employer of this description of labour, is the ale-can. But with these bad points there are many elements of the true pith and ring of the English character. Industry like that of the hive bee, sturdy toil such as that which was commanded by the builders of the pyramids, or the brick-building kings of Nineveh, firm fellowship and good feeling, evinced in subscriptions to sick funds and doctors' bills, clear-headed application of labour to produce a definite result; above all, a sense of the right that man and master alike have to fair play and honest dealing, all these virtues are found in the kit of the navvy. He is a man with whom there is some satisfaction in working, and a man as to whom you can attribute any failure in the attempt to elevate him into a position of permanent comfort and respectability, not to any inherent infirmity of nature, but to want of early training, and to the potent influence of strong drink.

It is otherwise with the Irish labourer, when in his own country. In England or America he becomes a hard-working and reliable operative. His natural ability, far above that of the average Englishman, when directed in the right channel, enables him to take a high rank in the republic of labour. At home this is not the case. His thoughts are bent, not to discover how to become a good workman, but to discover how to obtain the most money for the least labour. What may be the case when Irishmen work for Irishmen, it is not attempted to say. Let the Irish owners of land and employers of labour bear their own witness of their own experience. All that is here recorded is the result of an effort made by English Engineers to do in Ireland as they had done on this side of St. George's Channel: to benefit themselves, certainly, by the execution of important public works; but to benefit, at the same time, the districts in which they were laid out, and the labourers who were employed on them, by the fair and liberal method of remunerating labour.

The first point of trouble that arose, was one from which there was no subsequent escape. The Irish labourer, as the soil produces him, is totally and utterly untaught. A man who has in him the making of an excellent workman, but who has never handled a pick or a shovel, never wheeled a barrow, and never made a nearer approach to work than to turn over a potato-field with a clumsy hoe, is in no position to make a dock or a railway. He has to be taught, and be set to work. He can only be educated for a remunerative toil, by being organized by a competent foreman, and set, for a time, to work alongside of a steady workman. For this purpose a nucleus of English workmen is indispensable. It is not asserted that Northern Irish, or Scottish, labourers might not be substituted for English; but the question is out of the bounds of the writer's experience, and if it were otherwise, the result would be the same. The Irish labourer ranks himself by counties and by divisions of counties. The West Cork man would have more enmity against his fellow-countryman from Tipperary, or from another division of his own county, than against the unmitigated Englishman. Then, no sooner is work set out, gangs organized, foremen or small sub-contractors set to work, than the strong enmity of the mass of the local labourers is kindled. For a time it will smoulder, till a week or two's pay is   169 drawn, and then it will be sure to break out, (at least it always has done so,) in the form of strike, of menace, or of actual outrage.

Those men who are the life of any public works, men who will both toil with their own hands, and direct those who toil with them, by advice, as well as by example, are the first to catch the drops of the threatening storm. The manner in which it may break is as various as are the diversities of the atmospheric tempest. One time it is a letter addressed to the employer, and forwarded to him by that postal service which is native to Ireland—that is to say, by being nailed by a knife to the door of his office or his residence—that unless such and such men are removed from the works, steps will be taken to remove them. Sometimes the lives of the men are threatened; sometimes that of the employer; sometimes the milder course is adopted of a threat to leave the works without hands, and to prevent, forcibly, the arrival of men from any other district. Sometimes the evil consequences of maintaining the honest workmen are left vague, but none the less menacing from their shadowy nature.

Thus the first difficulty with which the English Engineer has to contend, is the opposition made by the organization of idleness to the organization of labour. It is one which alone must prove fatal, unless it is encountered not only by resolute and unflinching courage on the part of the master, but by the exercise of the same virtue on the part of a staff sufficiently numerous to carry out his orders. It is a case, as before stated, of the goose with the golden egg; for the men who join in a common cry against every successive foreman or sub-contractor, are themselves totally incapable of carrying on any skilled work. Employment must cease with the flight of the minor chiefs of labour. With employment, pay must come. Hundreds of men must return from the enjoyment of the grateful and unwonted luxuries of meat, and of beer, to potatoes and butter-milk. Yet conspire they always did, and, as far as the lesson of disagreeable experiences goes, they always will.

CHAPTER XXIV

Justice in Ireland

In those times when the use of armour made a very palpable difference between the security of the life of the knight and that of the peasant, a noble who unadvisedly strayed on the land of his feudal enemy, or even of a strange and unknown proprietor, was likely to be made to pay heavily for his folly. English history has this truth written in pretty legible character on its pages. The seizure of Richard Cœur de Lion by the Duke of Austria, cost the subjects of the Crusading King no trifling ransom; and the capture of the Saxon Harold by the Comte de Ponthieu led to the establishment of that Norman dynasty, which showed its title-deeds on the field of Hastings.

A similar act of folly, it may be thought by some persons, is committed by the English employer of labour, who finds himself, in the exercise of his vocation, on the soil of Ireland. He is fair game. He has to pay his ransom. It is the *lex loci*. Gentle and simple alike see the justice of the case, and unite in the resolution that the thoughtless invader shall not escape scot-free.

The permanent, persistent, conspiracy of idleness against industry, which has been one main feature of much of the trade unionism of this country, has been already referred to as springing native from the Irish soil. It is, no doubt, one of the most formidable influences with which the real friend of the labourer,—the large employer of labour—has to contend. Its fatal influence can be but feebly realized, except by those who have experienced its effects. But in speaking of the manner in which the less uneducated classes greet the efforts of the Engineer, it may be more instructive, and less invidious, to give instances of actual occurrence, than to indulge in general remarks.

When a man refuses to be terrified, and when it is difficult or inexpedient to murder him, there is a grand resource in Ireland to bring him to reason. England herself has furnished her ever-complaining sister with the means. England, in her simplicity, has offered her own cherished panacea—trial by jury. In her own case, this institution has proved a deterrent from a too ready resource to legal hostilities. A solemn appeal, as of old, to lot, or even the more modern

171

substitute for that ordeal, known as "tossing up," might perhaps render as substantial justice as an English common jury, and answer the same useful purpose of composing strife. But, in the hands of the Celtic population, the jury becomes altogether a different institution. It is no longer a matter of chance. Justice returns to preside over a trial, only it is not a justice dependent on the right or wrong of the case, but a justice depending on the local opinion of the district from which the jurymen are taken. When an Irish jury is empannelled, it is very often a sheer waste of time to enter into the question of evidence.

Three instances may be selected of the manner in which the Civil Engineer is brought, in the sister country, into a relationship with the dispensers and guardians of the law, which is unusual in English experience. A proceeding before the bench of Magistrates—a civil, and a criminal trial—are calculated to give more information by a simple recital than could be drawn from many an elaborate report.

One of the principal works on the railway, from the prosecution of which most of the incidents here described are taken, was a tunnel through a *grauwacke* hill, at some distance from any centre of habitation.[100] The trial shafts, according to the local Engineer, had been sunk without any difficulty from water. The fact that they resembled very excellent wells, when they were inspected by his English brother professional, the former attributed exclusively to the rain. On the principle that *all* springs are fed by rain, the Irish optimist was, no doubt, correct; but when serious work on these shafts was commenced, it became evident that the rain had percolated the whole hill, and that until powerful steam-pumping machinery was introduced, water was the chief material that filled the skips or buckets of the miners.

Around the shafts of this tunnel, as being the representative stations of the chief difficulties to be encountered by the constructors of the line, the elements of opposition collected with instinctive readiness. The division of the workmen into separate gangs for each shaft afforded the readiest method of indicating the whereabouts of any obnoxious ganger, or English navvy, against whose presence the patriotic feeling of the natives was excited. On one occasion a hardy sub-contractor engaged to perform the whole of the work for a definite measured price. He worked briskly for some weeks. Matters went on with unusual smoothness—progress was slow, but it was peaceful —and order seemed for once to reign above and under ground. One Friday afternoon, however, the sub-contractor having received and cashed the cheque for the balance of his fortnight's pay, took some other road than which led to his expectant workmen. Whither he went is immaterial, even if it were ascertainable; but whither he did not go was clear. It was not to the pay-table. The night passed

[100]The tunnel was through Goggins Hill, on the Cork and Bandon Railway.

172

in a scene to which it would require the pen of Lever to do justice; and early on the Saturday morning, a breathless messenger reached the bed-side of the Engineer in charge of the construction. This gentleman, who had suffered much in health, not only from the anxiety of his post, but from the dampness of the climate, had been ordered by his physician to refrain from any attention to business for a day or two. "They will burn the house over your head," said the pleasant messenger. "For God's sake run away, and let no one know where ye go, till the men are quiet again." But the other ordered his horse, and as soon as he could dress, galloped off to meet the three hundred men, who were on their march with hostile demonstrations.

The broadest street of a well-known Western city, where the offices of the Engineer were stationed, was full of these men. Robbed they certainly were. Angry, it was natural that they should be. The only thing wrong was, that they were bent on demanding redress from the wrong source. Giving up any effort to trace their own missing employer, they came to demand their wages of the Engineer who had set, not them, but him, to work.

The arrival of a blood horse in full gallop in their midst, the instinctive care of the noble animal on one side, and of the surging mass of labourers on the other, alone preventing a collision, was unexpected. Irishmen, too, like a display of pluck; and the idea of the man they sought, coming alone and, apparently, unarmed to meet them, was so unusual a phenomenon, that they met it with a stifled cheer. Improving the opportunity, the Engineer sent the crowd to a builder's yard in the suburbs of the city, where he promised to meet them in half an hour. The city Magistrates, meantime, met to express their fear of the consequences of the possession of one of the main thoroughfares of the city by so formidable a mob. But when once out of the streets, the worthy Magistrates left the parties to settle their own difficulties as best they could.

From about eleven in the morning, until the close of the short daylight of winter, the Engineer sat his horse in the midst of that extremely unpleasant circle. First, he heard all that they had to say, making them appoint two or three delegates to represent their views. Then he occupied them, by taking down the name and designation of every one present, a procedure which is usually regarded with alarm by a mob, and which often terrifies men who are disposed to lawless acts. At last matters came to a point. The men demanded payment. The Engineer admitted that they ought to be paid, but not by him; for it was not the mere amount of the two or three hundred pounds, to which the men were entitled from their employer: it was rather the question whether the Railway could or could not be made, that was in dispute. Had it once been established that, on the failure 173 or flight of their immediate employer, the labourers could demand arrears of wages of the constructors of the line, the principle of small contracts would have been destroyed, and the unlimited liability incurred by the chief contractors would have been so great, that no prices could have covered it.

The upshot of the long day, at any moment in which it was not clear that blood would not flow in the next, was, that the men returned to their homes, or their huts on the tunnel, unpaid and grumbling, indeed, but beaten in their attempt to fix the responsibility on the Engineer. Another sub-contractor promised immediate employment, and some advance of pay on account, to all who would at once return to work. The defaulter was to be pursued, and the men ultimately received their arrears. The fight was a hard one, but the result was not lost sight of.

Vengeance was attempted on an English foreman, who was assiduous and unremitting in his duty, a skilled man, who set out the time for the workmen, and measured off the work executed by the sub-contractors. The matter came to blows—several men were injured, and the life of the foreman was distinctly menaced. The matter was brought before the Magistrates. The Chairman of the Railway Company was the Chairman of the Board. The offenders were defended by an attorney, a functionary never wanting in Ireland when a dispute is in the wind.

The whole matter was deliberately investigated, and the man who had threatened the Englishman, was ordered for trial at the next assizes. The defence, so far, had been almost hopeless. Now came the tug of war. The attorney demanded that he should be bailed. The contractors strongly opposed this application. They felt that if the offender was allowed to leave the Court in the honourable position of a man under recognizances (as the labourers would regard it), that the life of no English workman would be worth a day's purchase. After a very long struggle, the Chairman of the Magistrates gravely announced that the Bench would admit the prisoner to bail. "I'll be — — if I ever set a foot on the works again," said the English foreman, throwing down an unloaded revolver on the floor of the Court, and making the best of his way in retreat.

"Mr. Engineer, will you favour me with one word?" said the Chairman of the Magistrates, leaving his chair for a minute. "Don't let your man be in such a hurry;"—added he in a whisper, "I didn't see that I could refuse to take bail, as a matter of principle, don't ye see, but divil take me if I don't reject any bail they offer. Make yerself asy; we'll not let him out of prison." No more they did. So it came to pass that the works were ultimately completed. But

it was not to be expected that Englishmen should wish for another

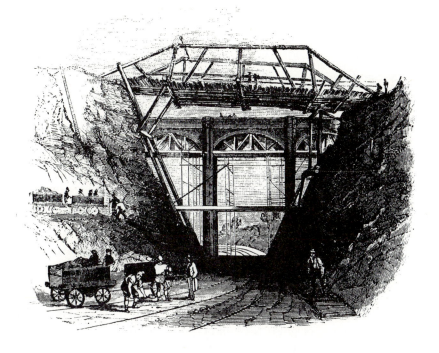

experience of this nature.

A second trial arose out of the same troublesome turmoil, earlier in its origin, but later in its settlement, than the case just mentioned. A foreman who had been employed, by the Engineer of the line, in sinking the shafts, and who was a witness as to their dry nature, was strongly recommended by that officer to the constructors, as a competent sub-contractor, to whom the execution of the tunnel might be safely entrusted. An agreement was accordingly entered into, by which the man, whom we may call by the name of Davy, undertook the excavation of the driftway, and the subsequent completion of the tunnel, at a given price per yard.

At first, as usual in such cases, all went on well; but as difficulties increased, and the only stable element in the case proved to be an ample and unfailing supply of water, Mr. Davy got into trouble, got into debt, and finally broke down. The constructors had to make new arrangements with fresh sub-contractors, a proceeding which, in such cases, always involves a loss. But when all was again in activity, Davy made the discovery that he had not been a sub-contractor at all, but an agent or foreman; and that, instead of having overdrawn his account, his masters were in debt to him, for his time and outlay, to the extent of three hundred pounds! This difference of opinion was referred to the lawyers to decide; and the case came on for trial at the autumn assizes for the county of Cork. Davy was his own witness. The principal evidence for the defence was that of an agent of the constructors, who spoke directly to the fact of the contract with Davy, and read extracts from his memorandum-book which, if they were not forgeries, could leave no doubt as to the trumped-up character of the claim. So clear was the case made by the summing-up of the judge, that on the conclusion of his Lordship's remarks, the attorney for the plaintiff said, half aloud, to the representative of the defendants, ''Deed, Mr. So-and-so, don't ye think ye were a fool for throwing away yer money to fee Counsel, when his Lordship has been your Counsel for nothing?'' But the upshot of the case showed that the plaintiff had not misjudged the sympathy of an Irish jury. The verdict was for the plaintiff, with a special addition, to the effect that the jury attached full credit to the evidence of the witness for the defence. The attempt at effecting a compromise between the financial, and the moral, result of the deliberations of the jury, would have been ludicrous, but for the light that it threw on the manner in which that Saxon institution works among a Celtic people.

A third instance of legal warfare over the corpse of this unlucky railway was yet more memorable, inasmuch as it engaged all the bar of the Western Circuit, and filled the newspapers with the details of a three days' trial. It was remarkable, also, as an instance of an able and unscrupulous resort to the criminal law, for the sake of obtaining possession of a property, not only without payment, but even without agreement as to a price.[101]

[101]The tussle described here prevented the opening of the railway on the advertised date in August 1851. See *The Times*, 13, 18, 19 and 22 August.

The English Engineers who had undertaken the completion of this Irish railway, received no money from the Company. The payment for works and plant was to be made in shares and debentures, and the Irish Engineer had given to his Directors a low estimate of the amount that would be required. As it became evident, in the course of construction, that this estimate was altogether inadequate, and as the financial powers of the Company appeared likely to be exhausted, the Engineer-in-Chief tried to put off the evil day of reckoning, by omitting to make the stipulated monthly measurement of the works, and declared that he would give no further certificates until the work was done. An unpleasant feeling was thus produced; but the constructors, instead of stopping the works, took the better course of pressing their conclusion with the utmost rapidity. So far had the line advanced towards completion, that the Engineer-in-Chief fixed a day for the opening, and advertised that trains would run from that date; and he placed police and inspectors on the line, and patrolled it by a locomotive, which he kept constantly moving over the rails.

The Engineer who had charge of the works on behalf of the constructors of the line, was at this juncture in England. But his staff thought proper to act in the emergency, with something of the vigour which their master was accustomed to show. They gave notice to the Directors that they should not allow them to take possession of the works, until they had been measured up and certified for; and as the preparations for the public opening were carried on without any attention being paid to this reasonable stipulation, they effectually blockaded a deep rock-cutting close to the terminus, by the simultaneous explosion of several mines in the slopes. On this the railway officials instantly took criminal proceedings, under Lord Campbell's Act for punishing obstructions to railways. The Assizes were close at hand; and as every member of the staff and establishment of the constructors was included in the indictment, not a single witness was forthcoming to rebut the evidence of the Engineers of the line.

The grand jury found a true bill without hesitation; and when the Engineer in charge of the construction, hastily summoned from London, arrived in Cork on the eve of the trial, he found all his people under detention as accused criminals, and the lawyers on his side looking very grave. The animus and moral character of the proceeding were plain; but the blow had been so rapidly and so ably delivered that it seemed likely to succeed. A condemnation of the defendants would have handed over the line to the Company without conditions.

The men now most eminent on the Irish bench and bar, were all engaged on the trial, either for the Crown, the Railway Company, or the defence:—the hereditary legal celebrity of Plunkett[102] and Sir Colman O'Loghlen;[103] the fervid eloquence of Butt,[104] that gifted but erratic genius who was a Doctor of Laws before he was out of his terms; the acumen, legal knowledge, and gentle bearing of Fitzgerald[105] and of Deasy,[106] now honoured members of the Irish bench. The first day was occupied by the case of the Railway Company, and at its close it seemed certain that they would win. They represented that, on the eve of opening the line to the public, when the trains had been arranged to run on the following morning, and when the officers of the Company were actually making use of an express train, a wanton and spiteful outrage on the part of the defendants had broken up the line, and endangered the lives of the officers in question. Cross-examination produced little or no result. The Engineer of the line, availing himself of the *legal* ignorance in which his refusal to measure the works had left him, denied any claim on the part of the constructors of the line. In fact the story, as told by one side without contradiction, was very black indeed.

When the case for the prosecution had been thus triumphantly

[102]John Span Plunket (1793-1871).

[103]Sir Colman O'Loghlen (1819-77).

[104]Isaac Butt (1813-79).

[105]John David Fitzgerald, later Lord Fitzgerald (1816-89).

[106]Richard Deasy (1812-83).

concluded, some dismay was caused to the advocates of the Railway Company by the appearance on the table of the sole witness for the defence:—the Engineer whom, from the fact of his having been on the other side of the Channel, it had been impossible to include in the indictment, that had shut the mouths of all other opponents of the Company's plaint. Objection was raised to the admission of this witness. It was natural that every effort should be made to exclude him, for his first reply produced a marked effect. "Are there any questions of account in dispute between you and the Company?" asked Mr Deasy. "The Company are in debt to us, on an account which they have refused to examine as they are bound to do by the contract, to the amount of more than thirty thousand pounds," was the reply; "and out of that sum, there is not more than four or five thousand pounds, as to which any possible question can arise." One of those little shudders, which evince the surprise of a crowded audience at some unexpected statement, ran through the Court. The keen Irish intelligence arrived in a moment at the real meaning of the trial. After a few more replies, the Judge[107] directed four or five of the accused persons to be immediately discharged. They were consequently available as witnesses as to the details urged by the prosecution. The case for the defence was complete; and the Judge, after entirely disposing of the question of any moral offence on the part of the remaining defendants, addressed himself with great care to the legal construction and purport of the Act of Parliament, under which this was the first prosecution. His Lordship concluded his charge by saying that, in cases where works were being carried on at a time that locomotives were being run, it had been shown in evidence that certain precautions were proper. In the present case these precautions had been taken, and they had been fully successful. No one had been injured or endangered in life or limb. With that he left it to the jury.

[107]Chief Baron Pigot (1797-1873).

It should be mentioned that, while it was admitted that the rock in question had been blown down with a view to obstruct the seizure of the line by the Directors, it came out in evidence, that the excavation thus effected had been ordered by the Engineer-in-Chief, and that, although effected in a summary and wholesale manner, it must have been completed sooner or later.

During the long deliberation of the jury, the solicitor for the defence accused himself of neglecting the interest of his clients. "There are two men from the other terminus of the line on the jury!" said he. "They will think it will hasten the opening of the line if they convict the prisoners. I ought to have challenged every man from that district, whoever he was."

178

The opposition, however, of these two local adherents of the prosecution was, after a time, overborne, and the verdict of "Not Guilty" was given in a crowded and breathless Court. The chief legal interest of the scene hinged on the attempt that was made to enlarge the scope of the operations of a new Act of Parliament. But the moral spectacle presented, by the appearance in Court of a public body, endeavouring to twist the criminal law into a means of ejecting the unpaid constructors from works which had been carried on at their own expense, and to brand, as convicts, respectable men, in order to avoid the settlement of an account, is one not easily to be forgotten. It showed that it was not the poor and the uneducated alone who were ready to have recourse to violence and to extortion. It showed that, if the cotter or the labourer thought that by strikes and by menaces he might secure pay without giving equivalent labour, men who ought to have set him a better example showed the same moral obliquity. It showed, in fact, that whatever temptations, in the way of promised gain, there might be for the investment of English capital in Irish works, there was a more than counterpoising disadvantage, arising from the totally different manner in which it is the English and the Irish habit to regard the question of right and wrong.

CHAPTER XXV

Engineering in the principality

We have heard a good deal, at one time or another, of representative men. Whatever truth, or whatever nonsense, may be uttered on the subject, it is certain that much interest attaches to what may be called representative facts. Circumstances may occur in which the influence of powerful causes, ordinarily evinced as working for a long time over a large area, may be concentrated, as it were, within a narrow space and in a brief period. Thus the results of a change, similar to that wrought in the more central parts of England by the operations inaugurated by Macadam and by Stephenson, as they were carried on for more than the third part of a century, were discernible, within three or four years, by those who had the opportunity of watching the introduction of the railway system into Wales.

Political, military, and commercial, communication between the capitals of the United Kingdom and of Ireland, led to the establishment of a postal communication through North Wales, which rendered the state of the country immediately served by the Holyhead Road an exception to the general condition of the Principality. The Suspension Bridge over the Menai Straits was one of the noblest public works of which England could boast. When St. George's Channel was bridged by steam, and when passengers were no longer liable to lie at the mercy of the wind in the barbarous little hamlet of Holyhead, the stream of traffic, if feeble, became constant. The influence exerted by the establishment of a mail route through a distant country district, can be but faintly imagined by those who have had no experience of the result, or at least have not been able to compare the state of opened, and unopened, regions of country, at some considerable distance from the busy centres of commerce and of intelligence.

North Wales had a second link which bound it to the activity of English civilization in the beauty of its scenery. While the regular service of the Holyhead mail enabled the tourist to count on a certain communication with Chester and the district of the East, the romantic landscapes found in the mountain systems of Snowdon, Plinlimmon, and Cader Idris, attracted numerous summer tourists. Thus North Wales, before it was traversed by the locomotive, stood in pretty close relation to the more densely peopled portions of the Island.

In Central and Southern Wales this was not the case. The once archiepiscopal city of St. David, formerly the seat of learning and the home of a University, was, and is at this moment, a miserable hamlet, with one small change house, unworthy to be dignified by the name of inn, or even of tavern. The roads were nominally mac-adamized, but for the most part they were either formed of that clay shale, blue or brown, which makes excellent footpaths, but which, beneath horse shoes and wheels, returns, in wet weather, (and it is mostly wet weather in Wales,) into simple clay mud; or were covered with unbroken pebbles, over which the horseman passed with care, and often with curses. The large undulations due to the geological character of the district, the constant climbing of toilsome and unshaded hills, the slow descent over the dangerous pebbly road, made travelling in many parts of Wales, even with the best appointed private carriage, a toil and a disgust. Much of the country presented no attraction, to overcome the force of this natural barrier to intercourse with England. In some districts the Welsh language alone is spoken, and Saxon-tongued strangers are not regarded with favour. Language and roads, race and habits, the natives, the climate, and much of the scenery, were anything but tempting to the tourist.

The Government service alone prevented the total isolation of the Southern Principality. At the extremity of South Wales lies one of the most superb harbours to be found on the coast of our Island, one in which the united navies of the world might, safely and conveniently, ride at anchor. Our national literature carries back the fame of Milford Haven to the period of the Roman occupation of Britain; and Shakespeare was not unaware of the natural grandeur of the unbuilt dock, to which he conducted the forlorn wanderings of Imogen. A small but active dockyard, famous for the excellence of its workmanship, had been established at Pater, the eastern horn or angle of Milford Haven, within a couple of miles of the county town of Pembroke. And as Dublin and the North of Ireland were reached from the Anglesea promontory, a postal steam service was established between Milford Haven and Waterford. A direct mail route was thus kept up, running over a fairly maintained turnpike-road from Gloucester to Carmarthen. The traveller had to climb to a considerable height, in order to pass the mountainous locality of Brecon, then skirted the romantic valley of the Usk, famous in the romances of King Arthur, and finally toiled, over barren wilds and steep hills, from Carmarthen to the point of embarkation.

This point was neither the town of Pembroke, which lies on a river at some distance from the harbour; the healthy and well-situated town of Milford, where the Government dockyard was in the first instance situated and whence it was driven, by the unwise rapacity of the landowners; nor even the newly-settled huts of Pater, or Pembroke Dock, the abode of the labourers in the dockyard. The ocean terminus of the Waterford mail was a single house of twelve or fourteen rooms, at once an hotel and a post-office. Close to this building, a fine limestone pier had been erected, by the aid of the diving-bell, in water so deep close in shore, that vessels could snugly lay alongside to take in coal, or to receive the diminutive mail-bag, and possibly the single passenger, brought down by the Carmarthen mail.

The River Cleddw, the union of the northern and southern branches of which forms the head of the Haven, is imperfectly navigable up to Haverfordwest, the county town of Pembrokeshire, where it is crossed by two limestone bridges. The sixteen miles of mountainous road, over which a branch mail, diverging from the track of the Waterford postal route, ran from a little hovel called the Roses, rendered this smart little town, which boasted four churches and the remains of a fine old castle, even more inaccessible than Pembroke, inasmuch as the roads were ill laid out and worse kept up.

About Haverfordwest, however, and on towards Milford, where a rich sandstone soil replaced the barren and impervious shale, or "rab", which renders the greater part of Pembrokeshire both unproductive to the farmers, and unwholesome to those who are unfavourably influenced by a damp atmosphere, is to be observed a singular ethnological phenomenon. You have passed for a long time through a Celtic population, and remarked wiry hair, snub noses, and prognathous jaws, which remind you of the county Kerry. English speech has stood you in but little stead. But you are surprised to find that, after a long passage, you have arrived at what the inhabitants call little England beyond Wales. A tall, well-formed, handsome, English-speaking, race,—civil, cleanly, and industrious, keeping themselves entirely clear from the Welsh, whom, sad to say, they are apt somewhat to hate and despise, and with whom they never intermarry, and but rarely hold any intercourse,—claim to be the descendants of Flemish settlers in this remote district. There is, perhaps, no part of the British Islands so productive of female beauty. The famous Nell Gwynn was a native of Laugharne, a little fishing village on the Cleddw between Haverfordwest and Milford Haven;[108] and representations of the large eyes, long deep lashes, and pure complexions, of many of the women who attend the weekly market at Haverfordwest, would be treasures in the portfolio of the painter.

The formation of the South Wales Railway brought this remote district into a relation with the busier parts of England, in a manner

[108]Much confusion here. Laugharne was not in Pembrokeshire at all but further east near Carmarthen, and Nell Gwynn seems to have been born either at Hereford or in London.

182

that was astonishing to the inhabitants. We have remarked that there existed a macadamized road; and the daily service of a mail was kept up summer and winter. But, in the latter season, it was not uncommon for the coach to be so delayed by the state of the roads as to have to start on its return journey to "the Roses", before the few letters which it brought down could be distributed, or, of course, answered. With the opening of the Great Western Railway to Gloucester, London became less inaccessible. The later and successive completion of the South Wales Railway, to Chepstow, to Swansea, and to Carmarthen,[109] added so many links to the chain that bound Milford to the Metropolis. But during the time thus occupied, in which works were commenced, suspended, and recommenced, in Pembrokeshire, the peculiar and picturesque character of the people was comparatively undisturbed by the great revolution which the locomotive was rapidly effecting, wherever the scream of the steam whistle, and the puff of the blast, ceased to be unfamiliar sounds.

[109]To Swansea in 1850, to Carmarthen in 1852.

The county magnates looked with disfavour on the railway. The main occupation of the Pembrokeshire landlords was fox-hunting. How could the hounds be expected to keep the scent, if people were allowed to make great ditches of twenty feet deep, and corresponding banks, across the country? The invasion was intolerable. Then as to wages. It was actually stated that the railway contractors had the extravagance to pay half-a-crown a day! (This was before the four-and-ninepenny pay to the navvy had been heard of so far west). Eightpence a day was fair agricultural remuneration. The Welsh labourer did not ask more, and if he should do so, he would not get it. It was immoral in the highest degree that English Engineers should come down to disturb this satisfactory state of things, and to make the country people discontented! So the county folk ensconced themselves in their dignity, and shot evil glances at the invading Engineers, if they chanced to meet them at church or at market.

The original idea of Mr. Brunel was to fix the terminus of the South Wales Railway at Fishguard, a little Welsh town on the shore of a bay of the same name, where the last debarkation of a hostile foot on British soil took place in 1795. On that occasion, the alarmed Welsh-women, attired in red cloaks and tall beaver hats, flocked together in such numbers, that the French invaders took them for the English army, and surrendered at discretion. A wilder spot for a terminus, as far as the surrounding country was concerned, could not have been selected; and the hydraulic works that would have been necessary to form an appropriate harbour in the exposed bay, would have involved a large outlay. The object, however, was to compete with the London and North Western, and the Chester and Holyhead, Railways, for the Dublin traffic. It was thought that the 183

superior speed attainable by the broad-gauge engines would be as marked over the sharp gradients of the South Wales, as over the level district of the valley of the Thames. From London to Reading a speed of 70 miles an hour was attained without difficulty. But when the curves became sharper, and the gradients rose to one in 200, or even to one in 100, the case was reversed. The establishment of a second gauge must be considered as, in some respects, an engineering mistake. On the other hand, it must be remembered that the original Liverpool and Manchester gauge of 4ft. 8½in., was arrived at by mere accident, and that it was unfavourably narrow. A six-feet gauge, or something a little under, would unite most of the elements of mechanical advantage. The additional foot claimed by the bold originality of Brunel, added more to the cost, than to the available working advantages, of so broad a gauge. The idea of the speed to be attained by ten-feet driving wheels, on the seven-feet gauge, was modified by subsequent experience. Yet there can be little doubt that the improvements in the narrow-gauge locomotives, and in the speed maintained on narrow-gauge lines, are due in no small degree to the imperious rivalry of the broad gauge. It can hardly be doubted that the broad gauge will ultimately disappear. It is unquestionable that a large sum of money will have been wasted by the temporary adoption of a second gauge; but it is possible that, in compensation, the nation at large owes much to the emulation between the advocates of the two systems.

Through a country like South Wales, however, it is manifestly desirable that a narrow gauge should have been adopted in the first instance. The natural features of the district are large, the gradients are consequently severe, and economy both in the works, and in the useless weight of the trains, was imperative. The minute care given by Mr. Brunel to the correction of the survey of the line, the careful cross-sectioning of all side-long ground; the shifting of the centre line so as to secure the minimum earthwork, failed to produce an economy equal to that which might have been effected by the use of the 4ft. 8½in. gauge. It was probably the tacit influence of similar considerations that led Mr. Brunel, on the occurrence of the check to railway enterprise that followed the political disturbances of the commencement of the year 1848, to abandon the idea of reaching Ireland by the Fishguard route. The Pembrokeshire works were for a time suspended; and on their resumption, in March 1851, the Cardiganshire terminus was definitely abandoned; and the line of the South Wales Railway was carried, first to Haverfordwest, to which a branch from the Fishguard line had been authorized, and then, under new Parliamentary powers, to Neyland, a point on the northern angle of Milford Haven, opposite to the Government dockyard;[110] and on the other side of the deep gut or channel, the southern shore of which was lined by the pier erected for the embarkation of the mails. A more admirable point for an ocean terminus it would not be easy to find. Shelter, and deep water, close in shore presented themselves at a minimum of expense. The works through the valley of the Cleddw, and indeed the whole forty miles from Carmarthen, were of considerable magnitude. Swing, or rather slide, bridges were erected over the Towy and the Cleddw, and the cuttings through the undulating surface of the ''rab'', or clay slate rocks, and through the solid red sandstone near Neyland, were heavy.

A feature unusual to be met with in public works was characteristic of Pembrokeshire. Plenty of stone offered itself at first sight to the Engineer, and the hard resistance offered by the soil to the pickaxe was a source of considerable expense. But when it was sought to extract stone fitted for building purposes, the case was different. Not only in the different railway cuttings, but in many promising localities, quarries were opened in search of building stone. The upper courses of the stone reached were of such grit and texture as to justify further outlay, the result of which was, however, in almost every instance, the abandonment of the quarry. The entire mass of stone had been so shaken, or fractured, by geological action, that it was impossible to extract sound pieces of a sufficient size for any

[110]The railway reached Haverfordwest in 1854 and Neyland ("New Milford") in 1856.

but the most paltry description of masonry.

The practical importance of a scientific acquaintance with geology was illustrated by the expensive failures met with in seeking to open quarries in Pembrokeshire. In the arrangement of mountain systems suggested by the French geologist, D'Orbigny, which are for the most part mainly characterized by their direction in Azimuth, the peculiarity of shattered structure is attributed to a single set of elevations. It is that of the Pembrokeshire district.

One fine bed of blue limestone, in solid, unbroken courses of from 12 to 24 inches, or even more, in depth, runs across the county. It is susceptible of the polish of marble; and its dark fossiliferous surface is an admirable material for chimney-pieces or table slabs. In the absence of ordinary stone, some of the Pembrokeshire bridges were constructed of this costly material. It was a remarkable contrast which some of these walls, of a rusticated masonry, almost cyclopean in its character, presented to others of the bridges near Carmarthen. The careful specifications, and the rigorous inspections, of Mr. Brunel, were overborne by the contractors in this district through the sheer weight of the imperfection of all material that was not fetched from a distance; and some of the finest, and some of the shabbiest, bridges, for which the Engineer of the Great Western ever certified, may be seen, within a mile or two of each other, erected at the same cost to the Company. The difference between working for profit and working for character is written, in large hand, amid the wilds of Carmarthenshire and of Pembrokeshire.

The future importance of Milford Haven may be safely prognosticated. It is a curious circumstance that if the globe be divided into two hemispheres, so that one contains the maximum, and the other the minimum, of habitable land, the central point of the former, or, allowing for the imperfection of chartography, at all events the central *port* of the former, is Milford Haven. Before the invention of the steam vessel the importance of this station, in a military point of view, was very great; as a fleet might put to sea from its shelter during the prevalence of winds which would close most other ports. Now that the introduction of the iron road has restored the old predominating importance of the most seaward points of departure, it seems certain that Milford Haven will become the great point of embarkation for the New World. Steamers, that would fear the sandbanks of the Mersey, may anchor or moor in safety in its sheltered water. It was proposed to Mr. Brunel that the Great Eastern should be built, and built afloat, at Neyland, a village being constructed for the abode of the necessary staff of workmen. After mature consideration Mr. Brunel declined the offer. He admitted its advantages in all respects but one, that of proximity to his office in Duke Street.

It was his intention to give so much of his personal attention to the details of this favourite offspring of his genius, that he determined, for that reason alone, on its construction on the Thames. Had he otherwise decided, not only would the great expense of the launch have been entirely avoided, and the whole cost of the giant steamer been greatly reduced, but a far greater, national, loss might have been averted. The Great Eastern, if built at Neyland, might have been completed at a less expenditure than that of the life of our greatest Engineer.

CHAPTER XXVI

Railway finance

So large a capital has been invested in the construction of railways, and so widely different has been the cost of that construction, reckoned per mile, in different instances, that a few words on estimate and account may not be without value.

The railways of Great Britain have been executed under almost every variety of condition. The two opposite and extreme cases of the conduct of public works may be thus regarded. The Projectors, or the executive officers of the Company, may execute them themselves by day labour; that is, the manager will engage all the workmen that he thinks fit, paying to each man his daily wages, and appointing competent inspectors to see that each man does his duty. In this case the check against the idleness or waste of the workman, lies entirely in the efficiency of the inspector. The individuals employed, unless they are inspired by a high and unusual sense of duty, have no inducement to exert themselves, but the fear of discharge, if they are too slack in their work. No wholesome emulation, either as regards quantity or quality of work, is encouraged in the mass of workmen. On the contrary, those who work hardest will often be reproached by their fellows for so doing, inasmuch as they thus furnish a gauge or standard by which the inspector can detect the insufficient quantity of the work of other men. Thus a large employer of labour, who pays exclusively by the day, enlists, to a great extent, human nature against himself.

This direct method of payment for time, which alone is available, or rather which alone has hitherto been employed, for military and naval purposes, is also that made use of in our dockyards, and generally in the employment of labour by the Government.

In the case of large works, the opposite plan has been not unfrequently adopted. A person comes forward, who undertakes, or contracts, to perform the entire work for a fixed and definite sum. All questions of employment of labour are therefore made over to this contractor. The employer makes a certain bargain. For so much money he is to have what he requires. Whether the actual cost be more or less is nothing to him. His opinion as to the adequacy of the price—of the lump sum, as it is called—may be formed either from professional knowledge or from the result of competition. He may invite a certain number of persons to tender, or he may call on any one who chooses to come forward, to do so. If he then accepts the lowest tender, and takes due security for the fulfilment of its terms, all care, trouble, and responsibility may be thought to be over.

Clear and simple as this may appear to be in theory, it is far otherwise in practice. In fact, a lump sum is a selling price. As such it is perfectly applicable to a simple commercial operation. So many tons of rails can be delivered, at a fixed place and date, at so much per ton. But even then it will be necessary, for those who want to have good rails, to have the manufacture inspected. But when the large variety of requirements needed for railway construction is considered, it becomes evident that a margin is requisite in every instance. The price of labour is only one item of cost. Land, timber, stone, bricks, iron, glass, every material used by the workman; the rent of lodging, the price of food, the state of the floating balance of the labour market, the total number of unemployed workmen available from time to time;—all these things have to be taken into consideration. Each item of work undertaken by a contractor bears its own risk. Every risk must be covered by a contingent profit.

Nor is this all. A railway cannot be made as a yard of tape can be cut off. It is requisite in the first instance to know exactly what is to be done. No doubt many a contractor will give a price per mile from a cursory inspection of a line of country; but prices thus roughly quoted imply either enormous profits, or the risk of the contractor's failure,—a circumstance always highly detrimental to his employers. Very careful plans, sections, and estimates have to be prepared, and it is further evident that it is only by a strict adherence to these details, on the part both of the contractor and of the Engineer, that this lump sum contract can be carried out in its integrity.

It is not in accordance either with human nature, or with the character of public works, that any such accurate adherence to plan should be maintained. Time is generally a material object; and the Engineer is, for the most part, too much pressed for time to have all his professional labour completed before the contracts are let. There is much that he will learn later. Even when the provision of boring to ascertain the nature of soil has been made, it often happens that the opening out of a large cutting, or tunnel, displays features in the contents, that were not indicated by the trial shafts. In every instance, as accuracy of information enables the Engineer to determine price with the smallest possible margin, vagueness and uncertainty cause an increase of estimate. Certitude of price is thus attained at the cost of increase 189 of amount.

This is not all. Alterations will be often required in the original design. Circumstances beyond the control of the Engineer may lead to a wide modification of his intentions. Improved means of carrying out his object may be discovered. The hands of the Engineer of a public work must be unfettered if he is to do his duty to his employers. Power to modify the contract must therefore be entrusted to the Engineer, and in case of any serious modification, the guarantee of the lump sum is at an end.

To meet, as far as possible, this difficulty, it is customary to append schedules to the contract. In the earliest contracts of Mr. Robert Stephenson for the London and Birmingham Railway, these schedules were three—one for the payment, on account, of contract work; one of additional prices for additional or "extra" works; and one of prices according to which deductions were to be made from the total amount of the contract, for any items contemplated in the tender, which the Engineer might direct to be omitted in execution. It became customary, later, to append a single schedule, applicable in each of the three described cases.

In either instance, however, it is impossible, by any provision, to replace or to supersede an honest and upright confidence between the Engineer and the contractor. When this subsists, each unexpected question will be arranged on its own merits. When it is absent, it is impossible to provide for unforeseen events which may place one or other of the contracting parties at the disposal of his opponent. The constant attention of Mr. Brunel was given for many years to the perfection of the contract specifications. The result of each successive new edition of the printed form was to place the contractor more and more at the absolute mercy of the Engineer. Bilateral rights were, as far as careful language could go, extinguished. The right to be paid, at a certain stipulated rate, for work ordered by the Engineer, was all that these documents were intended to leave to the contractor. If the former ordered nine-tenths of the work to be abandoned, the latter, who had prepared for the execution of the whole, had no redress. It was provided—and it was a wise and economical provision—that the contractor should be paid fortnightly, on the measurement of the Engineer, from 80 to 90 per cent. of the schedule value of the work executed. But even for this payment, on the regularity of which all his arrangements would depend, the contractor had no legal security. It was stated to be the intention of the Engineer to recommend a payment, instead of its being acknowledged to be his duty to certify the execution of a certain amount of work. So "tight" did these specifications ultimately become, that nothing but
implicit personal confidence in Mr. Brunel on the one hand, or the

belief that the Court of Chancery would set aside any unfair provisions, in case of dispute, on the other, could justify any thoughtful man in entering into a Great Western, or a South Wales, contract. During the latter part of the professional career of Mr. Brunel, after he had established that personal weight and prestige for which, at the commencement of the Great Western Railway, he had to make so determined a struggle, and before failing health and loss of nerve caused him to lose that absolute independence of action which, in his best days, he maintained with reference to the Directors of his various undertakings, it was neither unsafe nor unpleasant thus to execute his orders. Admit him to be absolute, and he was not only reasonable, but kind. Hint to him that you had rights, and he was inexorable. But this was an extreme case. It was tolerable only so long as the Engineer was uncontrolled, not only by the contract, but by the Directors. More than one man attributed, and rightly attributed, very heavy losses, to the fact of his having signed a rigid and unilateral contract, on the final settlement of which, Mr. Brunel found himself unable to make the allowances on which he had given the contractor the right to count. Equality before the law, entire submission to the Engineer on questions of science and of practical execution, with mutual confidence in skill, energy, and good intention, form the relation in which the contractor should stand to his employer, and which the contracts and specifications of the latter should be directed to establish.

The extreme cases of departure from the golden mean are, those in which the contractor attempts to rob the Company, and those in which the Engineer attempts to ruin the contractor. Instances of each extreme are, unfortunately, within the experience of most men acquainted with the progress of public works. On one occasion, a contractor brought in his bill to Mr. Robert Stephenson in so grossly exaggerated a form, that when, on his thankfully accepting some 80 or 90 per cent. of his claim as a settlement in full, the latter inquired, ''And how could you have the face to send me in such an account?'' he replied, ''Well, sir, I must own that I put on a topper for luck.''

An instance, on the other hand, of the manner in which the tables were turned on an Engineer who, either from incompetence or from ill-feeling, had cruelly scourged his contractors, is far from uninstructive. The work in question was a bridge, a portion of which was supported on heavy piles. The specification provided for a price per foot to cover all expenses of pile-driving up to, let us say, fifty feet—a ten-foot pile being at so much per foot, twenty-foot at so much, and so on. At from forty to fifty feet, a solid bed of gravel was attained, the superincumbent material being soft alluvial soil, so that by the entrance of the pile-shoe to its own depth into this solid   191 stratum, which rested in its turn on bog, the greatest possible amount of stability was attained. But the Engineer in question was not contented with this result. He had got into very hot water with the contractors and their agents, and he insisted on the use of sixty-foot piles, which could be provided only at a very great expense; and he ordered these piles to be driven ten feet into the ground. The result was, that many of the piles were shivered in driving, and had to be withdrawn, and that those which were sent through the gravel, offered less resistance than would have been the case if they had only entered it. Months were consumed instead of days, the work was, at least, far less satisfactory than might have been the case, and the certificate of the Engineer was for an amount less than a fourth of the total cost.

The contractors, annoyed at such a proceeding, and injured by a permanent weekly expenditure without return, consulted an Engineer familiar with contract accounts. He visited the spot, and carefully examined the contract. Then he called for the whole of the accounts from the commencement of the work. Observing that the schedule only gave a price per foot for piles driven up to fifty feet, and that it also included prices for timber, and for day labour, he charged for the whole of the sixty-foot piling as timber, and as day work, a mode of drawing up the account which was in exact accordance with the schedule, and which added three thousand pounds in one item to the contractor's bill. The right to make the charge in this form was indisputable, for it was the only way in which sixty-foot piles could be carried to account in the exact terms of the schedule. The Engineer of the bridge had outwitted himself, and his employers had to pay handsomely for his malevolent, or ignorant, caprice.

There can be no doubt of the enormous waste which has been incurred in the construction of the English Railways. For much of the outlay, the Engineers are scarcely responsible. Parliamentary and legal expenses represent a price per mile quite adequate to the construction of those light railways of which we are now so much in want.[111] Landowners have thriven largely at the cost of shareholders, first by exacting high prices and enormous residential and occupational damages, and then by finding the value of their whole property so much increased, that in most instances it would have paid them well to make a gratuitous cession of the land required, as an inducement to the Engineers to lay out the works through their property. Duplicates and opposition lines have been another source of profligate outlay. Mr. H. E. Bird, in a careful tabulated statement of the expenditure on 258 Railways, up to the close of June, 1867,[112] gives a total of £487,905,167 expended on 14,000 miles of line. It is highly instructive

[111]Conder had already referred to light railways on p. 3, pointing out quite truly that the power to construct such lines under the Regulation Act of 1868 was limited to existing companies and did not extend to the making of new railways intended from the first to be ''light''. It was consequently very little used in England until after 1896, when the law was changed. In France, Belgium, Holland, and Prussia, however, these railways were built extensively and proved both useful and profitable. Conder returns to this subject on p. 199.

[112]Henry Edward Bird (1830-1908) was a professional accountant, much interested in railways—although remembered chiefly as one of the outstanding English chess-players of his time. Conder is here using his *Railway accounts,* published in 1868.

192

to remark the different prices of English, Scotch, and Irish Railways as set forth in this statement: 9,634 miles in England have cost £405,331,055, or £42,000 per mile,—very nearly the price of the London and Birmingham Railway,—notwithstanding the immense economies introduced by subsequent experience; 2,466 miles in Scotland have cost £55,921,649, or £22,700 per mile; 1,898 miles in Ireland have cost £26,652,463, or £14,000 per mile. To say nothing of the low cost of Railways in the United States, which are far less solidly constructed than our own, but which have served for the development of a large and increasing traffic, at an aggregate cost, at the close of 1867, of 1,654,050,799 dols. for 38,605 completed miles out of a total of 54,325 miles undertaken, it is evident, from the comparison of work and cost in England and in Ireland, that our own waste has been enormous.

Had the English Railways been kept down to the cost of twice the Irish lines, a condition of which the neglect is due principally, or even exclusively to Parliament, our 9,634 miles, at a cost of £270,000,000, would be earning a gross revenue of £33,000,000 per annum, yielding a net return of from 6 to 7 per cent. on the capital. To speak of the difference of £135,000,000 as sheer waste will appear, to persons familiar with the subject, as an understatement.

That wasted outlay would have sufficed, to supplement our 9,634 miles of principal line with 45,000 miles of light branch railway, such as is now in full operation in Norway. The whole mileage of such a system would have nearly doubled that of the macadamized public roads. Great as is the impulse given to production and to commerce in England by the costly provision which we have made for the through traffic, the gain thus secured sinks into insignificance, in comparison to that which may be fairly estimated as likely to result from the application, to the iron roads, of that system of proportioning expense to duty, which regulates the outlay of the surveyor of the turnpike and the highway.

CHAPTER XXVII

The future of engineering

The profession of the Civil Engineer is under a cloud. Offices are closed, public works are discontinued, draughtsmen and pupils are unemployed. So great a depression in the condition of a liberal occupation, which, a few years ago, was only in want of men competent to follow it, is almost, or altogether, without precedent. It is a question of no slight importance, both to the Civil Engineer and to the public, whether this lack of employment results from a permanent, or from a temporary, cause. Have the Engineers of Great Britain sprung up like a mushroom growth, to disappear, having produced the fruit of the railway system, as rapidly as they were called into activity, or have they yet a social function to discharge, of which such men as Watt, and Brindley and Stephenson, have been only the pioneers?

To answer this question, it is necessary to examine the cause of the present state of affairs: causes, rather than cause, it may be said; for political reasons, and influences beyond the bounds of Great Britain, are not foreign to the subject. But as far as the Engineer himself is concerned, there can be no question that the main cause of his present inactivity is overproduction. He has done too much. He has crowded the work of half a century into a few years. There is little blame to be attached to him on the score of the quality of his work. He has covered the island with noble viaducts and lofty embankments, furrowed it with excavations, and pierced it with tunnels. He has found the means of supporting the passage of enormous weights, at high velocities, over lofty and wide openings, and has bound England firmly together by an iron network. But he has too often done this in anticipation of the wants of the day. He has provided, by a large present outlay, for a traffic that is undeveloped, and, what is worse, has provided often in duplicate. He has lent his aid to the litigant and to the speculator. He has produced much, which, if good in itself, is not good in its conditions. He has not remembered that a sum, laid out in public works which produce no remunerative return, is, within a certain number of years, not only actually wasted, but more than wasted, inasmuch as it has reduced the productive area of the country, and entailed annual loss as well as annual expense. He has, to some extent, furnished the arms, by which overgrown speculation has robbed the public, without even permanently enriching the speculators.

But, while this is undeniable, too much of the blame must not be laid upon the Engineer. We must not overlook the tendency of the ever-present instincts of human nature. Where is the barrister who will at once advise his client that he has not a leg to stand on, and had better do anything than go to law? How numerous are the physicians who will tell their patient, that temperance in diet, in labour, and in relaxation, and careful attention to the laws of health, will do far more for them than prescription or than drug? Where are the clergy who hint that it is possible to over-build in church, or in chapel, accommodation? Each man looks with a natural favour to his own employment, and the *esprit de corps,* or *de metier,* joins with the honest desire for a full employment of individual industry, to lead him to do so, more or less, whenever he is consulted.

Then it should be borne in mind that the Engineer, like the lawyer or the doctor, is generally consulted on a special case. He is not called on to advise generally. He may be asked to estimate the cost of a certain work, or even to aid in the comparison of estimated profit, with probable outlay. He can hardly be expected to decline a service of this kind; it is not at all clear that he ought to do so, for the reason that another set of persons are proposing a rival scheme. He may be of opinion that his own is the best, or that either of them is practicable or advantageous. He cannot hold that both are susceptible of contemporaneous execution. His duty to his clients will lead him to do his best for their scheme, and if a duplicate project is improperly sanctioned, the onus lies, not on the advocates of either competing plan, but on the legislature, on which the duty of protecting the public theoretically devolves.

While, then, it is possible that, in many instances, an improvident and reckless expenditure has been not only unopposed, but, to some extent, carried on, by the Civil Engineer, it is not fair to lay at his door the *lâches* of the House of Commons. The entire want of any statesmanlike principle applied to the regulation of railway enterprise —nay more, the entire want of any moral principle so applied—is not the disgrace of the Engineer. He has not unfrequently been placed in the position of an advocate for an unworthy client, and it has been, moreover, for the most part, before an unjust, an interested, or a capricious tribunal, that he has had to plead.

The speculators who have made use of the talents of the Engineer, to the present injury of his profession, and to the misfortune of the country, have been chiefly of three classes. There have been 195 the Directors of existing railways, who have entered into a reckless contest for custom, which they might have commanded by a careful and provident use of means within their grasp. There have been the builders and contractors of railways, who asked for more of these works in order that they might have the profit of making them. And there have been the speculators, "pure and simple"—the gamblers in scrip and shares—who sought for an Act of Parliament, not that they might earn a fair return from the execution of a public work, but that they might make a profit on the sale of shares at a premium. This last class of gamblers has done more to injure public enterprise, than both the others put together. Without their ready aid, no questionable and unremunerative undertaking could ever have been launched. To their eager greediness, the inflation of such periods as 1845 was chiefly due. And when they had done the utmost mischief as bulls, on the turn of the tide they became equally formidable as bears. The depreciation of an undertaking can be organized even more surely than its launch. Its intrinsic value is equally lost sight of in either case. It is not railways alone, or even chiefly, that have experienced this dishonest usage; bank after bank, in the great crash of 1866, was brought down, not by its inherent weakness, but by the conspiracy of those who thrive on the ruin of unsuspecting victims.

The other misleaders of the Engineers have, for the most part, paid the penalty. The plutocracy of contractors has, to a great extent, vanished.[113] Princely dwellings, purchased by the toil of the navvy, have again become tenantless. Great names have swelled the columns of the "Gazette." And the lesson that that honesty which looks not only at individual accumulation, but at the permanent benefit of the country, is the best policy, has been written in the very largest text-hand over the doors of Great George-street.[114]

As for the Directors, they have escaped, for the most part, with that modified contrition which attends those who make ducks and drakes of the property of other people. The sorrow, no doubt, is intense, but somehow it does not turn the hair grey. It does not lay hold on the imagination with the same force that a much smaller personal loss invariably excites. So that the real penalty of the over-production is divided between the shareholders without dividends, and the Engineers without work.

Yet Great Britain owes no trifling debt to the Civil Engineer. If we compare the map of the island in 1760, with that issued by Bradshaw in 1868, and if we compare the description of the social state of the inhabitants at the former time, with the present activity of life, we shall be struck by the magnitude of the revolution to which the century has given birth. Political causes in France led to that memorable and terrible revolution with which modern history rings,

[113]Here Conder spoke too soon. It is true that Peto and E. L. Betts had gone bankrupt in 1866, and Thomas Savin in Wales. But Brassey weathered that storm and was still active, the biggest man of them all (he left well over £3 million when he died in 1870); so were Joseph Firbank, George Wythes and Charles Waring.

[114]i.e. the headquarters of the Institution of Civil Engineers, 25 Great George Street, Westminster.

but the difference in the physical and social condition of the remote provinces of France, that has followed the fall of the old *régime*, is less marked than that which has attended the opening of the means of intercommunication throughout England. For the packhorse, and for the slow and cumbrous waggon, Brindley taught us to substitute the canal-boat; and Macadam enabled us to attain a speed of ten, of eleven, and of sixteen miles an hour, by the "Quicksilver" mail, and the Shrewsbury "Hirondelle." Three thousand miles of canal, at a cost of fifty millions sterling, Mr. Smiles tells us, brought water communication to within fifteen miles of any part of England. That was the service rendered to the country by the Duke of Bridgewater and Mr. Brindley. The importance of Mr. Macadam's invention was even greater. The state of our roads in 1745 is now almost incredible. The march of the Pretender on London, seems just to have directed the attention of the Government to the exigencies of military roads.

But the point which is most surprising is that of the decadence of the country in this vital element of national welfare, since the time of the Roman occupation. We find traces of the handiwork of that great Engineering race, who were accustomed to fortify their camp even for the bivouac of a night, and to connect their more permanent points of occupation by roads drawn as the crow flies. It was not the habit of ancient writers to record, except incidentally, those features of social importance of which we are now accustomed to write such detailed descriptions; and it may be hard to discover, with absolute certainty, the exact antiquity of the excellent Breccia roads of Italy.[115] It is certain, however, that they are not attributable to English examples. With regard to that more expensive and more permanent form of road-making, which we have as yet been unable to dispense with in our great city thoroughfares, the pitched roads, the fragments of the Via Appia that exist in the neighbourhood of

[115] *Breccia* means broken stone or gravel, much the same material as the McAdams had used in Britain.

197

Puteoli and other places, and the long buried streets of Pompeii, show that Italian road engineering has been identical in its character for two thousand years. A mile-stone of Trajan, from which our English mile-stones might have been copied, still stands on the route which Horace took from Rome to Brundusium, and it seems rather to have been the want of that important invention which we owe to the steel-workers of Sheffield—the application of springs to wheel carriages—than to the state of the roads, that travelling in Italy, in the Augustine age, was not more distinctly in advance of travelling in England, in the time of the accession of the House of Hanover. A level, metalled road is of comparatively little use for speedy traffic, without springed vehicles. On the other hand, springs would have been useless for vehicles plying on the savage tracks, that afforded our internal means of communication before the time of Macadam.

What our fathers and our grandfathers have seen to be effected in the way of national progress, by the introduction of canals and of turnpike roads, our own eyes have seen exceeded by the invention of iron railways. In actual speed, the locomotive has not so far distanced the best appointed four-in-hand coach, as that vehicle outstripped its predecessors—the waggon, the packhorse, or the riding poster. But when cheapness, speed, facility for the carriage of weight, and punctuality, independent of weather or seasons of the year, are considered, it may well be thought that we have not yet realized the full advantages of the use of steam as a tractive power. Ten times the capital expended in canals has been employed in the construction of a network of railways of not quite five times the mileage: an outlay, in round numbers, of £500,000,000 on 14,000 miles of line. It is at the close of this outlay that the Engineer, and his great employer, the public, have paused to take breath.

If we compare the development now attained by the railway system, with that of our earlier means of land and water communication, if we bear in mind that the price which we have paid for the construction of each mile of railway in England, would have paid for nearly two miles in Scotland, for exactly three miles in Ireland, and for five miles in America; if we remember that during the period of railway construction, from 1833 to 1863, our imports and exports, which had been almost stationary, had quintupled—we are led to the conclusion that there is yet much, very much, work awaiting the skill of the Engineer in the construction of iron trackways alone. It is very clear that many of the 9,700 miles of English railways must have been made at an unnecessarily extravagant cost. The wise provision of proportioning outlay to requirement has been lost sight of. Not only so, but the prudence of the Engineer is frustrated by the

standing orders of Parliament, and he is forbidden to construct branch lines of that light description, and for that moderate price, that would enable him to carry into the more remote rural districts, the immeasurable benefit of direct communication with the centres of consumption by unbroken railway communication. Our English railways have cost £42,000 per mile. Branch lines may be grafted on the system at a cost, taking foreign experience as a basis, of £3,000 per mile, or a fourteenth of our actual mean outlay. That the country would find so large an economy in the use of railway branches, constructed at anything approaching to such a minimum price, as to demand the prompt completion of such a supplement to the more costly trunks and arteries, there can be little doubt, if the injudicious and detrimental exactions of Parliament were set aside. When we compare the free scope given to unnecessary and unwarranted competition which, by encouraging rival lines, has made the service of certain districts involve an outlay of from £80,000 to £130,000 per *useful* mile, with the prohibition to use a light railway for the service of a country district, or for the extension of the convenience of railway omnibuses to cities and towns, it must seem that the legislators of 1868 are as actively hostile to the true welfare of the country, as have been the ignorant obstructors who, ever since we began to drain and embank our fens, have endeavoured to stifle every improvement in the cradle. Opposition to scientific engineering is not less disgraceful when it takes the form of the resolution of a Parliamentary Committee, than when it attacks turnpike gates under the name of Rebecca,[116] or designates the cutter of a canal as the enemy of the agriculture of the country.

But while it is, at the present moment, chiefly the barbarous state of legislation that interferes with the natural and beneficial progress and completion of the railway system, it is not as a road-maker alone that the Civil Engineer should regard himself, or seek to be regarded. We must remember that in this country, and in this century, the most eminent Engineers have taken up an entirely new position. They have taken the command of the forlorn hope of human progress. Our earliest road-makers, canal-makers, bridge-builders, have been, no doubt, original, but theirs has, for the most part, been the originality of ignorance. All honour to the men—we owe them no less; and their names may stand deservedly higher in the roll of fame, from the fact that they had to invent, and did invent, what more highly educated and better informed men would only have had occasion to copy. Our canal-makers might have served an apprenticeship in Holland to the craft in which they were self-taught. Macadam's invention would have been long anticipated by any surveyor, who had been sent to report on certain Italian roads. These men, inventors and benefactors as they were, by their own strong original sense,

[116]These were riots of protest against the tolls levied by turnpike trusts in South Wales in 1842-43. The leaders were often disguised as women, claiming that they and their followers were Rebecca and her children, coming to "possess the gate" of their enemies (Genesis, xxiv, 60).

199

carried out the indications which nature herself afforded, with no unstinting hand. It was otherwise with Watt, with Stephenson, with Brunel. New powers were brought into energy; Nature, wrestled with hand to hand and year after year, yielded deeper secrets to these men than lay on the surface of her dominions. In all the mighty progress due to the discovery of the expansive force of heat, educated invention and well-informed thought have been brought to bear. Without ceasing to be practical, the Engineer has become scientific. Still triumphing over physical difficulty, he has done so by mastering the laws on which that difficulty depends. He has substituted for the rule of thumb, the organized application of principle.

Regarded in this light, it is to the Civil Engineer that the care of the physical advance of the human race will be naturally committed. The economy of labour will be his charge; a political economy more true and more beneficent than that of the lecture-table, the pamphlet, or the hustings. Certain great questions, which have been neglected until their neglect involves not only cruel waste but imminent peril, are now demanding solution. The increasing pressure of a population that doubles itself in a century, the increasing difficulty of maintaining health, or even life, in cities that double their population within forty years, calls for the thorough organization of that service which may be compared to the circulating and the digestive systems of animal life. The water supply of cities and towns, pure, ample, and efficient in case of fire; the removal and disposition of sewage; the redemption of our rivers and brooks from a neglect that is rapidly converting them into pestilent sewers; the application to agriculture of that mass of chemical fertilising power with which we now poison rivers and estuaries; the drainage of land, both main and subsidiary ; the storing-up of that precious water, of which we are either anxious hastily to get rid, or helplessly destitute; and the production of fertile and certain crops by irrigation; all these are but so many features of one department of duty of the Engineer:—the proper distribution and utilisation of the rainfall.

The application of Engineering to agriculture is another department of the profession which is yet in its infancy. The steam plough has already established its claim to rank as the best servant of the farmer. Portable and convenient steam-engines are fast superseding the more costly labour both of man and of beast. Our fields, when prepared for seed, are more like the beds of a well-tended garden, in many instances, than the rough, though picturesque, grounds into which our fathers cast their corn, or their turnip-seed. But the farmer is still, to a most unnecessary extent, dependent on the changes of a very variable climate. When his crops are ready to house, the obedient service of the steam-engine is not as well adapted as might readily be the case. Hay grows mouldy; wheat shoots in the stack, if an unseasonable set of showers intervene at the time of housing or of reaping. The Engineer is aware that this is entirely unnecessary. He can readily find the means, if consulted, of ingathering crops, independently of the weather, and of drying hay and corn independently of the sun. By the application before referred to, of scientific forethought and practical ability to the due distribution of the rain-fall, whether it be normal, in excess, or in defect, by the mechanical culture of the soil, by the use of the steam-engine in harvesting, in storing, and, when needful, in drying his crops, the greater number of the evil chances that beset the farmer will be provided against, and the food of man and beast will be no longer left at the mercy of the elements in a climate of proverbial uncertainty.

These are not mere imaginations or theories. They are nothing but the application of the teaching of experience. What can be done by improved means of communication, by draining, by irrigation, by the application of sewage manure, by steam culture, and by steam harvesting and drying, has been shown in individual instances. What will be the result, on the national welfare, of the general application of these known improvements, it is not easy to state with exactitude, although some idea may be formed by any one who can work a rule-of-three sum. There is a busy and a useful future for the Engineer who takes this view of his duties, and who finds himself engaged by those who do the same. That the vegetable produce of England might be doubled by the services of the Engineer, is a proposition not very hazardous to maintain.

Nor is the productive capacity of our own island alone, the limit of this part of the duty and service of the Engineer. The connection between production and distribution is of the most intimate nature. Production must, of course, precede, but the stimulus to that production on the one hand, and its available results on the other, depend closely on the means of distribution. The land occupied by roads is as necessary to the feeding of mankind, as is the land which nourishes corn, or supports cattle on its pasture. This land is to be found, more or less available, wherever the foot of man can tread. The perpendicular limestone cliff, where a basketful of soil is painfully lodged in a crevice, becomes rich with the fruit of the olive. The wide plains of the Antipodes pasture sheep, the flesh of which is actually thrown to waste, while it would be worth from nine to fourteen pence a pound in London. In the mechanical and chemical preparation of the food which is now lost, and in its economical transmission to the centres of consumption, is to be sought a method of lightening the burden of life, for the artizans of the crowded districts of civilization. The general problem of producing, in the most appropriate 201 and ready spot, the great requisites of human consumption, of ingathering, storing, protecting, and transmitting them to the proper market, is one which will not always be left to private enterprise, or to uninstructed commercial speculation. The superintendence of the most economical production and transmission of the raw material of human food, and human clothing, is part of the economy of labour, the field and function of the Engineer.

It is, therefore, not as undervaluing the result of the labours of the other working bees of humanity,—the security afforded by a competent military organization, the humanising influence of the teacher, the comprehensive prevision of the statesman (if a statesman should again arise in England), the self-devoted toil of the medical man, that we consider the profession of the Civil Engineer to be that on which, if it is rightly considered and followed out, the progress of humanity and of civilization, at least for the remainder of the present century, must principally depend. The sword, and the pen, and the lancet will be needed so long as war, and ignorance, and disease prevail on earth. The instrument by which these evils can be most efficiently checked—the instrument that would be held in honour, if the earlier weapons of violence and of oppression were superseded, is the spade. When we first hear of the introduction of man upon earth, when earth was described as a garden, the occupation of the noblest terrestrial form was to dress it and to keep it.

Index